PHASE2 ADVANTAGE

Your Cybersecurity Career Starts Here®

INTRODUCTION TO

Intelligence Analysis and Cyber Threat Modeling

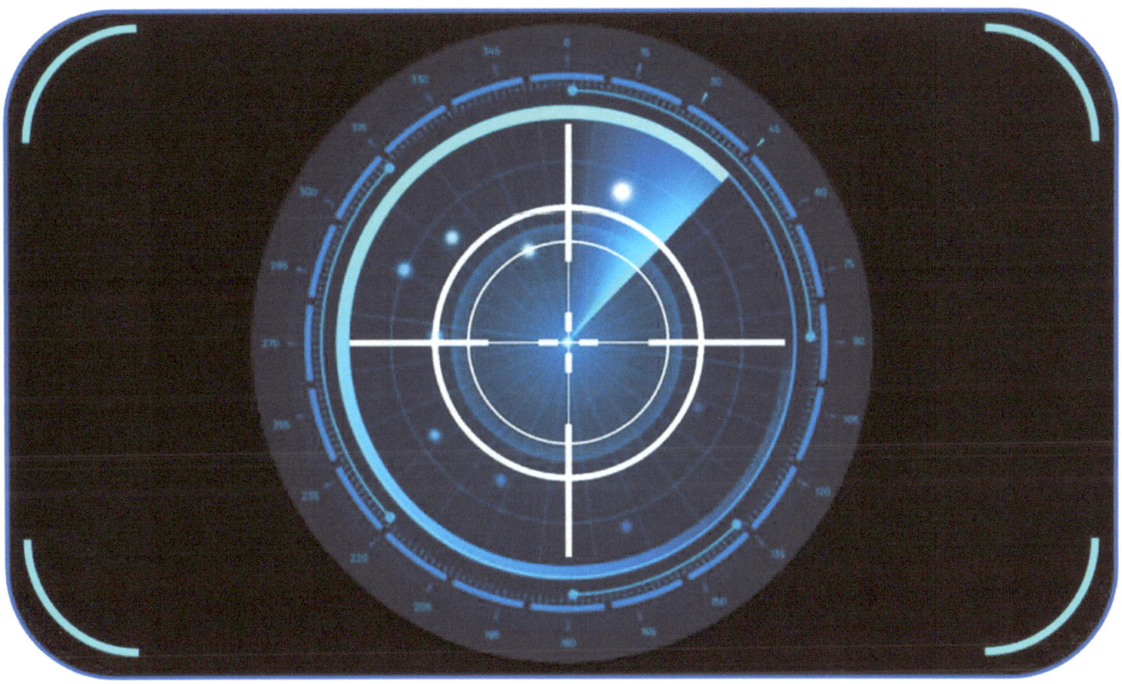

COURSE WORKBOOK AND STUDY GUIDE

Michael I. Kaplan

AUTHOR

01

READER FEEDBACK

Phase2 Advantage strives to publish cybersecurity textbooks of the highest standards and quality. Each text is subjected to a thorough development, editing, and review process prior to publication. This process continues throughout the life cycle of the publication.

Reader feedback is a valuable component of this process after the textbook has been published. To submit comments regarding the improvement of the quality of this textbook, contact us via email at *editor@phase2advantage.com*. Please ensure to include the textbook title and ISBN in your email.

Introduction to Intelligence Analysis and Cyber Threat Modeling
Course Workbook and Study Guide

Series: *Introduction to Cybersecurity*

Phase2 Advantage
P.O. Box 14071
Savannah, GA 31416
(912) 335-2217

Phase2 Advantage textbooks and immersive lab subscriptions may be purchased at discounted rates for educational, business, or governmental use when bulk orders are required. For information, please contact our office.

ISBN-13: 979-8987468241

COURSE DESCRIPTION

Threat Intelligence Analysis allows analysts to gain tactical, operational, and strategic advantages for customers of their intelligence product. Using highly developed analytical skills, analysts can maximize the benefits and efficiency of their supporting technology and automation, and significantly increase the security posture of their organization.

The *Introduction to Intelligence Analysis and Cyber Threat Modeling* course presents military grade concepts that are currently available only in intelligence schools and government agencies. Students will be provided with knowledge of core principles such as the Target-Centric Intelligence Life Cycle, Analytical Skill Development, Collection Management, Targeting Models, Structured Analytic Techniques, and considerations for creating high-performing Intelligence Teams.

The target audience for this course includes cybersecurity personnel serving in roles such as SOC Analysts, Law Enforcement Investigators, Red and Blue Team Members, and Threat/Warning Intelligence Analysts.

ABOUT THE AUTHOR

Michael I. Kaplan is the Director of Operations for Phase2 Advantage, a cybersecurity training and publishing company based in Savannah, Georgia. He is also the Chairman of the Savannah Technical College Cybersecurity Advisory Committee and heavily involved in curriculum design initiatives.

Michael has written numerous courses and cybersecurity training programs for corporate, academic, and government personnel. He has also developed training programs for Law Enforcement and Fugitive Task Force Investigators on the topics of Criminal Topology, Forensic Document Analysis, and Investigations.

Michael's technical areas of specialization are Threat Intelligence Analysis, Incident Handling and Response, Network Forensics, Digital Forensics, and Information Technology Risk Management. He also provides consulting services for government, corporate, and academic organizations both domestically and internationally.

ADA ACCESSIBLE COURSE MATERIALS

All Phase2 Advantage digital course materials – including textbooks, lab guides, and lecture slides in PDF and PPT formats – are ADA accessible and score **100%** on major Learning Management Systems such as Moodle, Blackboard, Canvas, and LearnUpon. For more information, please visit the Phase2 Advantage website at phase2advantage/higher-education.

Introduction to Cybersecurity Defense and Operations

The *Introduction to Cybersecurity Defense and Operations* textbook addresses the critical topics and skills needed to ensure a robust security posture is maintained. Readers will be presented with knowledge of core foundational principles such as Threat Intelligence Analysis, Vulnerability Management, Biometric Systems, Incident Response, Securing Systems with Cryptography, and implementation strategies of the NICE Cybersecurity Workforce Framework.

Paperback: 342 Pages

Publisher: Phase2 Advantage

ISBN-13: 979-8987468234

Introduction to Network Cyber Defense and Investigations

The *Introduction to Network Cyber Defense and Investigations* textbook addresses the critical topics and skills needed to ensure organizational networks remain secure and functional for its users. Readers will be presented with knowledge of core foundational principles such as Enterprise Network Devices, Network Services, Host and Network Forensic Analysis, Evidence Collection, Chain of Custody Requirements, and Reporting and Documentation requirements.

Paperback: 324 Pages

Publisher: Phase2 Advantage

ISBN-13: 979-8987468210

Introduction to Cyber Risk Management and Recovery Planning

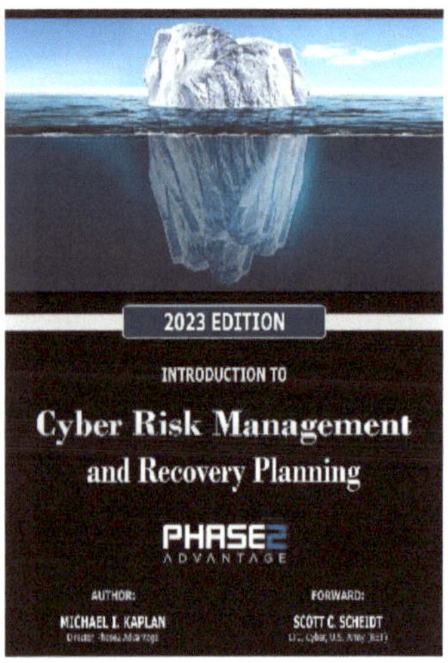

The *Introduction to Cyber Risk Management and Recovery Planning* textbook addresses the critical topics and skills needed to ensure continuity of operations under adverse conditions. Readers will be presented with knowledge of core foundational principles such as Risk Management Frameworks, Qualitative and Quantitative Analysis, Asset Inventories and Resource Profiles, BCP design, Cloud Computing Agreements, Cloud Security, and Legal Requirements related to Cloud Storage Solutions.

Paperback: 288 Pages

Publisher: Phase2 Advantage

ISBN-13: 979-8987468203

Table of Contents

Chapter Lecture Summaries

CHAPTER 01

The Purpose and Value of Intelligence

Lecture Summary

KEY KNOWLEDGE POINTS

Describing Intelligence by Type

The Primary Mission: Decision Advantage

Common Analytical Errors

Intelligence Sources and Methods

Strategic, Operational, and Tactical Intelligence

The Limitations of Intelligence

Intelligence Operations

Definition:

The variety of intelligence and counterintelligence tasks that are carried out by various organizations, and activities within the intelligence process including planning and direction; collection; analysis and production; processing and exploitation; dissemination; and evaluation and feedback.

Dictionary of Military and Associated Terms
jcs.mil/Portals/36/Documents/Doctrine/pubs/dictionary.pdf

The Intelligence Analyst

Definition:

An individual that reviews information from multiple sources, disciplines, and perspectives in support of intelligence planning and operations; evaluates and synthesizes large quantities of data (which may be fragmented and contradictory) into high quality, fused targeting and intelligence products; recognizes and mitigates cognitive biases and factual errors which may affect the analysis process; and recommends analytic approaches or solutions to problems and situations for which information is incomplete or for which no precedent exists.

NICE Cybersecurity Workforce Framework
niccs.cisa.gov/workforce-development/cyber-security-workforce-framework

Describing Intelligence by Type

Intelligence is often described in terms of the methods used in its **collection**, the gathering of various types of data through a variety of sources such as those listed below.

1. **HUMINT** (Human Intelligence): *Human intelligence* is intelligence gathered by means of interpersonal contact, a category of intelligence derived from information collected and provided by human sources.

2. **COMINT** (Communications Intelligence): *Communications intelligence* is information gathered from the communications of individuals, including telephone conversations, text messages and various types of online interactions.

3. **GEOINT** (Geospatial Intelligence): *Geospatial intelligence* is the analysis and visual representation of security related activities on the earth. It is produced through an integration of imagery, imagery intelligence, and geospatial information.

4. **SIGINT** (Signals Intelligence): *Signals intelligence* can be defined as a category of intelligence comprising either individually or in combination all communications intelligence (COMINT), electronic intelligence (ELINT), and foreign instrumentation signals intelligence (FISINT), however transmitted; or more simply as intelligence derived from communications, electronic, and instrumentation signals.

5. **OSINT** (Open-Source Intelligence): *Open-source* intelligence is information that is publicly available, appearing in print or electronic form including radio, television, newspapers, journals, the Internet, commercial databases, and videos, graphics, drawings, and textbooks. Most of the information collected by agencies and organizations in intelligence operations is open-source, and usually available to all who seek it out.

Intelligence can also be further defined in terms of means other than the collection methods used to acquire information.

1. **Analytic Process**: Any of the processes used to convert raw information or intelligence into a finished intelligence product.

2. **Implementation Level**: This will include tactical, operational, and strategic intelligence.

3. **Mission Type**: This will include examples such as indication and warning, projection, and deception and/or denial.

4. **Category Type**: This refers to the industry utilizing the intelligence such as military, business (competitive intelligence), political (opposition research), and cybersecurity.

The Primary Mission: Decision Advantage

Decision advantage is the ability to find out an adversary's intentions, and then enabling action to preempt, prevent, frustrate, or disrupt those plans. This is the primary mission for analysts tasked with providing their customers with an intelligence product. This can prove to be a challenging task. If successful, it can provide access to an adversary's *decision cycle*, which then has the potential to provide the benefits listed below.

1. The ability to alter the adversary's decision-making process.

2. The ability to shift the adversary's preferred outcomes.

3. The ability to alter the adversary's established target models.

Decision Cycle: The OODA Loop

The organization or adversary who has the most decision advantage is the one who can cycle through the *OODA Loop* the fastest and do it correctly. The OODA Loop is a recognitional decision-making model developed by Colonel John Boyd (USAF) and consists of the components listed below.

1. **Observe**: Have visibility into the target environment.

2. **Orient**: Ensure all elements of the target environment are properly contextualized.

3. **Decide**: Choose to execute a decision with intentionality, understanding that no action is considered to be an action unto itself.

4. **Act**: Once a decision has been made to execute, do so without hesitation. Evaluate the impact of the action through observation – the first step of the decision cycle – and repeat the process continually as needed.

Sample Analysis Scenario

This analysis scenario is designed to have students think through possible solutions using logical reasoning skills. The logical reasoning skills used to evaluate this scenario are equally as important as the answer. The directions for this scenario are listed below.

1. **Situation**: Two men are standing in the middle of a room facing in opposite directions; one facing East, the other facing West. The man facing East is 6'4", the man facing West is 5'08." The men are not allowed to move at all. The room is 10'x10'x10' and well-lit by a lighting fixture in the ceiling. The walls and ceiling are white. There is a 12"x12" mirror on the South wall, and a 12"x12" window on the North wall. The two men in the room cannot see the mirror or window in their peripheral vision, nor can they see any reflections cast by the mirror, window, or lighting fixture.

2. **IR (Information Requirement) Tasking Order**: Determine what actions need to be taken that will allow the two men described in the scenario to be able to see each other.

3. **Rules of Engagement**: No actions may be taken to alter the environment or any of the components presented in this scenario. Nothing may be added, removed, or manipulated by participants in this analysis. All facts needed have been presented; there is nothing "hidden."

Final Analysis: Intelligence Product

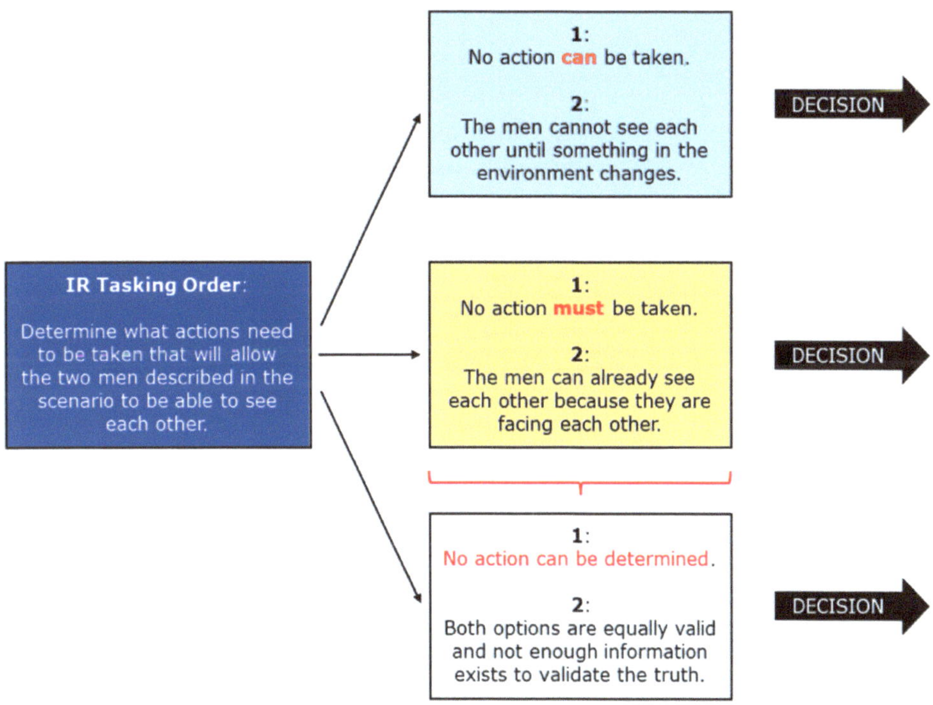

There are three possible outcomes that an analyst can present to a customer in response to this IR (information requirement) tasking order.

1. *No action **can** be taken*. The men cannot see each other until something in the environment changes.

2. *No action **must** be taken*. The men can already see each other because they are facing each other.

3. *No action can be* ***determined***. Both options are equally valid and neither can be chosen until more information becomes available.

The purpose of this scenario is to determine if the student can visualize all possible outcomes without becoming fixated on a single choice. Typical goals of intelligence are to establish facts and develop precise, reliable, and valid inferences for use in strategic decision making and operational planning. Failure to identify all possible outcomes is one of the common analytical errors listed below. The bolded text represents concepts which will be discussed in detail later in the course.

1. Improper **framing** of the situational context.

2. Improper **issue decomposition** to identify sub-issues.

3. Failure to identify **alternative competing hypotheses** (ACHs).

4. Failure to apply the principle of ***Occam's Razor***.

5. Facilitating **confirmation bias** with irrelevant data.

6. Limiting the analysis with **premature closure** due to lack of options.

7. Failure to separate facts from **opinions and assumptions** during the verbal portion of the IR Tasking Order.

Intelligence Sources and Methods

Intelligence operations are often considered *asymmetric* activities, allowing weaker organizations to compensate for their shortcomings and vulnerabilities. The sources and methods used in intelligence operations are as important as the information they provide. Their use tends to support activities that further the strategic interests of the organization. Although improved outcomes are not guaranteed, any effective intelligence they generate can prove to be a force multiplier. Sources and methods fall into the two categories listed below.

1. The people and tolls used to **collect information**.

2. The people and tools used to **conduct operations**.

Strategic Intelligence

Strategic intelligence is best understood in terms of the methodology used in most strategic planning known as *SWOT* (**S**trengths, **W**eaknesses, **O**pportunities, and **T**hreats). This type of intelligence deals with long-range issues and tends to be produced for the senior leadership of the organization. Examples of standard uses of this level of intelligence include strategic planning and budgets, technology development plans, and implementations designed to increase the overall security posture of the organization.

Operational Intelligence

Operational intelligence is used by national-level and international business organizations to support and plan for preemptive cybersecurity programs and focuses on the capabilities and intentions of both known adversaries and unknown potential adversaries.

Customers requiring an operational intelligence function typically prefer *predictive intelligence*. This level of intelligence helps the organization model the adversary's *tools, tactics and procedures* (TTPs), Assess the impacts of existing topology and assets on the adversary's capabilities, and monitor changes to the adversary's tactics based on a given environment.

Intelligence analysts soon discover that predicting future actions is difficult under any condition. Complete information is always lacking and not always accurate. In those instances in which complete information is available, the predictive intelligence provided to the customer may be accurate during early stages but change as the adversary adjusts to the customer's reaction to the intelligence.

Finally, the collected information believed to be accurate may be part of a *denial and deception* campaign as part of the adversary's tactics.

The six primary concerns of operational intelligence are listed below.

1. *Detecting* threat activity
2. *Investigating* threat activity
3. *Identifying* threat activity
4. *Targeting* threat activity
5. *Disrupting* threat activity
6. *Defeating* threat activity

Tactical Intelligence

Tactical intelligence refers to quick-reaction intelligence that supports ongoing operations by identifying immediate opportunities and threats (SW**OT**). This type of intelligence is used in the front line of any conflict and is driven by the need for fast response. Also known as *current intelligence*, it is used every day in a wide variety of situations.

Tactical intelligence that is properly implemented can contribute the success of investigations, such as those encountered during incident response operations and internal violations of an organization's acceptable use policies. It can also provide analysts with functional damage assessments which give visibility into the organization's ability to disrupt the target's (adversary's) effectiveness.

The Limitations of Intelligence

Although intelligence at all levels – tactical, operational, and strategic - provides a great advantage in supporting organizational interests, there are significant limitations inherent to it as well. This may prove to be frustrating to analysts but it is a reality which must be accepted. Several examples of these limitations are listed below.

1. *Some things are simply unknowable*. Analysts always want to have all the facts to inform correct decisions, but this is rarely possible. Information gaps must be filled with the analyst's best estimates, and these are never absolute.
2. *Variables in accurate forecasts are dynamic*. An analyst may provide a customer with an intelligence forecast that is 100% accurate at the time it was issued, but variables within the forecast are dynamic and subject to change without notice.

3. *Relevant variables are difficult to discern*. Analytical variables that appear to be relevant to analysts may be unimportant and dismissed variables may be critical. It is difficult to discern all relevant variables in the analytic process, especially when the analyst is operating from incomplete information.

4. *Adversaries may react to forecasts*. Analysts may provide accurate forecasts to customers who then act based on the intelligence provided. Adversaries may react to the customer's actions and adjust their decision cycle, therefor preventing the forecasted outcome from becoming a reality.

5. *Unforeseen forces will always emerge*. These forces can take the form of actors or circumstances (natural occurrences or human interventions) and are very difficult to foresee. In most cases, these forces present as unpleasant surprises and may significantly alter the original intelligence product.

CHAPTER 01

The Purpose and Value of Intelligence

Knowledge Assessment Questions

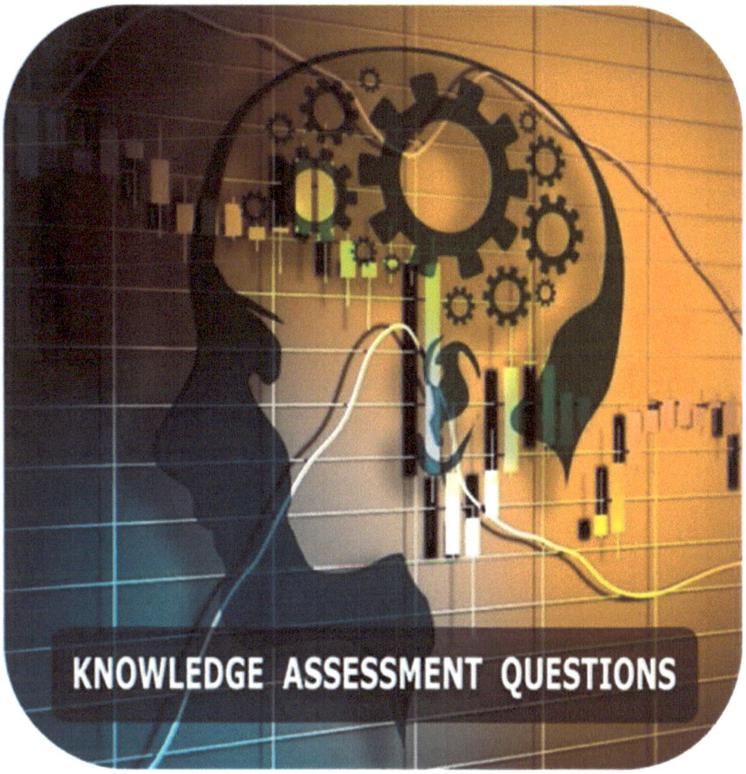

NOTICE TO STUDENTS:

The following knowledge assessment questions are presented in true / false, multiple choice, and fill-in-the-blank formats. The correct answers are provided in an Answer Key at the end of Chapter 14. These questions may or may not be presented on the quizzes and/or certification exam given by the instructor of this course.

Knowledge Assessment Questions

1) Describing intelligence by type can be defined by implementation levels such as __________, operational, and strategic.

A. Tactical

B. Mission

C. Technical

D. Asymmetric

2) The *OODA Loop*, developed by Colonel John Boyd, is a recognitional decision-making model that includes the actions observe, orient, decide, and __________ as part of the cycle.

A. Assess

B. Recognize

C. Analyze

D. Act

3) What choice below matches the following definition: "*Intelligence-gathering by interception of signals, whether communications between people or from electronic signals not directly used in communication*."

A. HUMINT

B. COMINT

C. SIGINT

D. GEOINT

4) A SWOT analysis is a technique used to help an organization identify __________, weaknesses, opportunities, and threats related to strategic intelligence and operational planning.

A. Situations

B. Similarities

C. Strengths

D. Symptoms

5) Which of the choices listed below would be a goal of an analyst engaged in an operational intelligence action against a threat activity?

A. Interrogating

B. Targeting

C. Dissuading

D. Directing

True / False

6) Modern intelligence activities involve the planning and coordination of multiple organizational units and are executed by teams, not individuals.

A. True

B. False

7) Most of the information collected by agencies and organizations in intelligence operations is classified, and usually available to all who seek it out.

A. True

B. False

8) Although intelligence provides a great advantage in supporting organizational interests, there are significant limitations inherent to it as well.

A. True

B. False

9) Strategic intelligence refers to quick-reaction intelligence that supports ongoing operations by identifying immediate opportunities and threats.

A. True

B. False

10) Intelligence operations are often considered asymmetric activities, allowing weaker organizations to compensate for their shortcomings and vulnerabilities.

A. True

B. False

CHAPTER 02

The Intelligence Process and Life Cycle

Lecture Summary

KEY KNOWLEDGE POINTS

The Intelligence Process

The Intelligence Life Cycle (Traditional)

The Intelligence Life Cycle (Target-Centric)

The Target as a Network

The Team-Generated Intelligence Model

Benefits of Target-Centric Intelligence

Chapter 02: *The Intelligence Process and Life Cycle*

The Intelligence Process

The analyst's primary job is to develop a level of understanding of the target and communicate that knowledge to the customer of the intelligence product. Intelligence is always concerned with *targets*, the focus of the problems or issues needing answers. The goal for analysts is determining the best way to approach the target. Developing an understanding and communicating knowledge typically consists of the four steps listed below.

1. Determine the steps needed to solve the intelligence problem.
2. Use a target-centric approach supported by relevant methodologies.
3. Communicate any knowledge and understanding gained to the customer.
4. Communicate in a manner that allows the customers to properly act on that understanding.

Although most of the focus is on collection and operations, most of the failures in intelligence are due to inadequate or nonexistent analysis. The job of an analyst is challenging and can be made more so by several factors. Large volumes of information are available but most of it tends to be either misleading or irrelevant. The irrelevant information must be examined and discarded - a task that takes much of the analyst's time - and it is easy for analysts to become oversaturated with incoming information. As a result of this information overload, critical data may not be shared effectively with those who need it.

Analysis Scenario: A Risky Jump for the Family

This analysis scenario is designed to have students think through possible solutions using logical reasoning skills. The logical reasoning skills used to evaluate this scenario are equally as important as the answer. The directions for this scenario are listed below.

1. **Situation**: Who would jump out of an aircraft without a parachute? If you agree to take the risk, you will be paid $10 million USD - *if* you survive. If you don't, the $10 million USD is guaranteed to be paid to your family, ensuring financial stability for the rest of their lives.

   An experienced skydiving instructor will jump with you and will have an extra parachute (for you, and it works *perfectly*). You can ask for it any time before or during the jump. The instructor will give it to you and secure it properly *without hesitation*.

   However, if you ask for (and accept) the parachute, you - and your family - will *immediately* lose the opportunity to receive the $10 million USD. On the bright side, both you and your family will remain safe and healthy.

2. **IR (Information Requirement) Tasking Order**: Based on the facts, would you be willing to selflessly risk your health and well-being to guarantee the future financial stability of you and your family?

3. **Rules of Engagement**: No actions may be taken to alter the environment or any of the components presented in this scenario. Nothing may be added, removed, or manipulated by participants in this analysis. All facts have been presented; there is nothing "hidden." However, there may be questions needed for clarification. If so, ask your course instructor.

Final Analysis: A Risky Jump

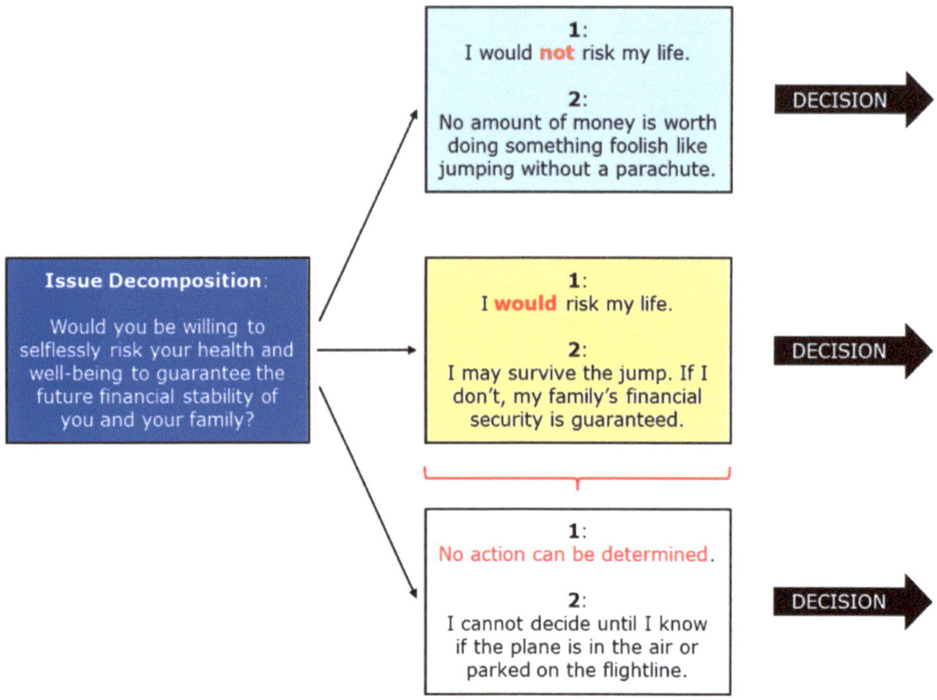

There are three possible outcomes that an analyst can present to a customer in response to this IR (information requirement) tasking order.

1. *I would **not** risk my life*. No amount of money is worth doing something foolish like jumping out of an airplane without a parachute.

2. *I **would** risk my life*. I may survive the jump. If I don't, my family's financial security is guaranteed.

3. *No action can be **determined***. I cannot decide until I know whether the airplane is in the air or parked on the runway.

The purpose of this scenario is to determine if the student can visualize all possible outcomes without becoming fixated on a single choice. Typical goals of intelligence are to establish facts and develop precise, reliable, and valid inferences for use in strategic decision making and operational planning.

The Intelligence Life Cycle (Traditional)

To have an intelligence capability, a process must be developed to take raw information and produce a product that can be used for action. The traditional Threat Intelligence Life Cycle provides this capability using the six phases listed below.

1. Planning and Direction

2. Intelligence Collection

3. Intelligence Processing

4. Analysis and Production

5. Dissemination of Intelligence

6. Utilization and Feedback

Planning and Direction

The *Planning and Direction* phase is composed of all activities that identify information requirements and ensure requirements are met. It manages the end-to-end intelligence cycle. Information requirements (IR's) are prioritized, collection plans are prepared, and collection capabilities are tasked.

The requirements of this phase are based on leadership (the consumer of the intelligence product). The IR's are managed as the situation changes so that collection, production, and dissemination efforts are utilized in the most efficient manner.

Intelligence Collection

The *Intelligence Collection* phase acquires and provisions the information to be provided to teams in the next chain of the intelligence cycle. Individual teams are allowed to collect their specific portions of the IR's which contributes to a redundancy of information. Redundancy allows assessments to be confirmed or disproved and facilitates cross-communication between teams. Once the collected information is reconciled it can be forwarded to the next phase of the intelligence cycle.

Intelligence Processing

During the *Intelligence Processing* phase, processed information regarded as actual intelligence will be mapped to correlating points for future phases. This phase includes exploitation of the data and a conversion process which may include decryption, translation, and interpretation. The information is now regarded as actual intelligence and is organized in a format specific to the organization.

Analysis and Production

The *Analysis and Production* phase analyzes, evaluates, interprets, and integrates raw intelligence into finished intelligence products. This final product must be delivered in a timely manner and focused on the needs of the stakeholder. The stakeholder decides if the intelligence report is complete, accurate, and if additional collection efforts may be required. This report provides an overview of the subject area in a clear, concise manner that allows the stakeholder to draw conclusions.

Dissemination of Intelligence

Intelligence Dissemination portrays an accurate picture of the threat to leadership in a form that facilitates rapid understanding of the issue. Listed below are several examples of the many types of architectures utilized for intelligence dissemination.

1. *Supply-Push*: No request is needed.

2. *Demand-Pull*: Intelligence is provided as needed.

3. *Standard-Channel*: Routine intelligence is provided to customers
4. *Alarm-Channel*: Critical intelligence is provided to customers.
5. *Broadcast Mode*: All authorized users receive intelligence simultaneously.
6. *Point-to-Point Mode*: Intelligence is only sent to specific users.

Challenges to the Traditional Intelligence Life Cycle Model

The traditional model may adequately describe the structure and function of the intelligence *community*, but it does not describe the intelligence *process*. Effective intelligence efforts are not true cycles; they are continuously evolving feedback loops within each other. This model defines an antisocial series of steps in which processes are moved from stage to stage, when in actuality the analysts in the process are continuously interacting. From an antisocial perspective, the responsibility for the final intelligence product can be avoided.

The traditional model also separates analysts from the collectors and treats customers as abstract entities. In this reality, true cycle feedback loops are absent. Finally, the rigid nature of this model makes it predictable and highly vulnerable to the adversary's countermeasures.

The Intelligence Life Cycle (Target-Centric)

An alternative to the traditional intelligence life cycle is to make all stakeholders, including customers, an intrinsic part of the intelligence process. This is accomplished with the target-centric model that supports a network-centric collaboration process. The primary goal of this model is to construct a shared *target picture* for all participants involved in the process, and it demonstrates the four characteristics listed below.

1. The process can handle complex problems and issues.
2. The model takes advantage of information technology.
3. All participants in the process are encouraged to contribute their knowledge.
4. The process does not follow linear sequences or traditional cycles.

The Team-Generated Intelligence Model

In the target-centric process, collectors and customers can not only see into the process but have many more opportunities to contribute to it and support the analyst's efforts. It is much more interactive than the traditional life cycle and is designed to encourage interaction among members. Because all participants can share knowledge of the target, it presents a better was to handle complex intelligence issues.

When all participants collectively share their knowledge of the target, all team resources are focused on the target. In this environment, participants are better able to identify and share gaps in knowledge regarding the target. They are also able to better identify and understand critical issues impacting the target and the environment in which it resides.

The Target as a Network

The goal of target-centric intelligence is to develop knowledge of the adversary's network, make the analyst's network more effective, and to render the adversary's network ineffective. In the traditional model, systems are viewed as individual targets. This narrow focus downplays connections within the network that may provide valuable information to the analyst. Connections can be real strengths for the adversary in the form of support and execution. Likewise, connections can present weaknesses for the adversary (communication and financial transactions). The points of interest within a target network are its *nodes* and *links*.

1. **Nodes**: Nodes can be people, places, things, or concepts.

2. **Links**: A link defines the relationship between nodes.

Components of an Effective Intelligence Network

It may be easier to see the adversary's side as a network than to see that one's own intelligence assets form a network and to fully exploit its strengths. Creating and maintaining flexible and empowered intelligence networks is difficult but worth the effort. These types of networks typically share the characteristics listed below.

1. Robust communications and connectivity within the team.

2. The ability to leverage physical and cultural proximity to support missions.

3. A network comprised of members with a shared vision and purpose.

4. Formal, established, and accepted decision-making processes.

5. Professional and personal relationships inside and outside the network.

6. Members with the knowledge and authority to effectively decide and act.

Object-Based Production (Target-Centric)

Object-based production is being implemented as a whole-of-community concept that fundamentally changes the way information and intelligence is organized. It is an intentional process method that organizes intelligence efforts around *objects* (targets) and features cloud-based sharing of knowledge.

The object becomes a *bucket* (container) used to store information, serving two valuable purposes for the team. First, the bucket becomes a single point of convergence for analysts needing information regarding the object. Second, it serves as an initial starting point for the intelligence effort, avoiding duplicate efforts and starting from scratch each time a new tasking order is issued.

Spatial and Temporal Target Attributes

In addition to being a system and a network, targets of intelligence interest have *spatial* attributes and *temporal* attributes that must be considered by analysts. Many intelligence targets are fixed geographically, such as elements of a region's infrastructure (datacenters or power grids). However, many targets are mobile – at least intermittently – and are characterized by features in both space and time.

In these cases, teams must use tools at their disposal to identify and locate targets of interest. This can take the form of tracking internet users, devices using geolocating IP addresses, or features that are hardwired into the technology device itself.

Benefits of Target-Centric Intelligence

The analyst's job is to take information from disparate sources and provide support to many customers with different requirements and needs. The target-centric model supports this mission well for the following reasons:

1. Analysis can take many forms.
2. Analysis can draw on many sources.
3. Several types of sources are often utilized for analysis.
4. Multiple sources create a more complete picture for the analyst.
5. Extrinsic sources which are often utilized can be effectively documented.
6. Analysis is always a continuous and complex process.
7. Process functions are always ongoing and must always be managed.

CHAPTER 02

The Intelligence Process and Life Cycle

Knowledge Assessment Questions

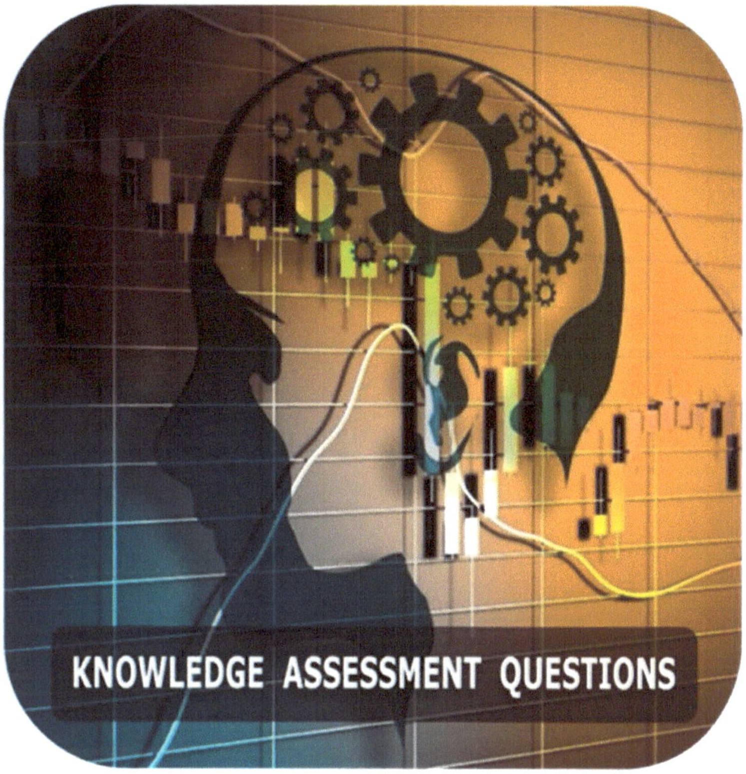

NOTICE TO STUDENTS:

The following knowledge assessment questions are presented in true / false, multiple choice, and fill-in-the-blank formats. The correct answers are provided in an Answer Key at the end of Chapter 14. These questions may or may not be presented on the quizzes and/or certification exam given by the instructor of this course.

Knowledge Assessment Questions

1) The __________ phase of the intelligence cycle is comprised of all activities that identify information requirements and ensure requirements are met.

A. Analysis and Production

B. Planning and Direction

C. Utilization and Feedback

D. Intelligence Processing

2) In the __________ intelligence life cycle process, collectors and customers can not only see into the process but have many more opportunities to contribute to it.

A. Traditional

B. Customer-Centric

C. Target-Centric

D. Community-Centric

3) In the network-centric intelligence model, systems are viewed as individual targets and networks are comprised of __________ and links.

A. Connections

B. Cells

C. Edges

D. Nodes

4) Object-based production in the target-centric intelligence model organizes the analyst's efforts around objects (targets) and features a __________ sharing of knowledge.

A. Cloud-Based

B. Clearance-Based

C. Security-Based

D. Agency-Based

5) The __________ phase acquires and provisions information to be provided to teams in the next chain of the traditional intelligence cycle.

A. Intelligence Processing

B. Utilization and Feedback

C. Planning and Direction

D. Intelligence Collection

True / False

6) Although most of the focus is on collection and operations, most of the failures in intelligence are due to inadequate or nonexistent funding.

A. True

B. False

7) The goal of intelligence is to develop knowledge of the adversary network, make the analyst's network effective, and to render the adversary network ineffective.

A. True

B. False

8) In addition to being a system and a network, targets of intelligence interest have special attributes and technical attributes that must be considered by analysts.

A. True

B. False

9) It may be easier to see the adversary's side as a network than to see that one's own intelligence assets form a network and to fully exploit its strengths.

A. True

B. False

10) In the traditional intelligence process, collectors and customers can not only see into the process but have many more opportunities to contribute to it.

A. True

B. False

CHAPTER 03

Adversaries in the Modern Threat Landscape

Lecture Summary

KEY KNOWLEDGE POINTS

Cyber Adversaries vs. Cyber Defenders

Nation-State Threat Actors

Advanced Persistent Threats

Non-State Threat Actors

Legal Challenges in Intelligence Collection

Respected Threat Intelligence Sources

Chapter 03: *Adversaries in the Modern Threat Landscape*

Cyber Adversaries vs. Cyber Defenders

Malicious actors – our cyber adversaries – will always have some degree of an advantage over cybersecurity professionals tasked with protecting the network. A side-by-side comparison of the two opposing sides should make the advantages and disadvantages obvious. Although several disadvantages faced by cyber defenders pose significant challenges, they are not insurmountable.

Cyber Adversaries	**Cyber Defenders**
No Regard for Laws	Hack Back Prohibited
No Limits or Boundaries	Regulation and Compliance
Tax Free Funding	Limited Budgets and Personnel
Acquire Exploit Technology	Change and Version Control
Unlimited Time Frame	Prove Security Value Daily
Must be Right Once	Must be Right Always

Nation-State Actors

The role of *nation-state actors* in cyberattacks that involve hacking and data theft has gained substantial attention over the last two decades. They have the power to exert influence over citizens and compel them to act as agents—willing or unwilling—and provide indirect economic support to malicious actors to conceal official involvement in illegal activities. Listed below are three criteria that must be met to meet the definition of a nation-state actor.

1. **Sovereignty**: Possess the authority to rule and govern.
2. **Territoriality**: Possess recognized borders and boundaries.
3. **Abstraction**: Possess an independent persona from the population it governs.

Advanced Persistent Threats (APTs)

An *Advanced Persistent Threat* (APT) is a threat actor, typically a nation-state or state-sponsored group, attempting to gain unauthorized access to a network. The actor's motivations tend to be political or economic and the targeted sectors include all industries. The threat actors are highly skilled and knowledgeable; they have specific objectives, capability, intent, and access to intelligence. Multiple targeting methods are often combined and executed with coordinated human actions.

8 Unique Characteristics of APTs

Many common attack vectors were initially introduced as part of an APT campaign using zero-day exploits, malware, and customized credential theft. Traditional cybersecurity measures such as defense-in-depth, firewalls. and antivirus cannot protect against a sophisticated and well-planned APT attack. The eight characteristics listed below distinguish them from less sophisticated actors and explain why APTs are leaders in the field of exploits.

1. Sophisticated actors

2. Specific objectives and methods
3. Substantial financial resources
4. Extended time of operation
5. Intelligence resources
6. High risk tolerance
7. Origin of attacks
8. Value of attacks

Nonstate Actors

Nonstate actors may work individually or complete attacks with a wide network of participants and then disband rapidly. The actors tend to have fewer resources, their attacks are typically less sophisticated, and they generally profit from data theft. Although they may have different targets and methods, they still have the potential to disrupt the operational capabilities of their victims and cause fear among the general population. They also benefit from a higher degree of flexibility as they are not restricted by traditional command-and-control hierarchies.

Components of the Modern Threat Landscape

The Internet and computer technologies have dramatically reshaped the threats posed to modern society and the ways we defend against them. Listed below are several crimes associated with this landscape which have either emerged or increased exponentially over the last two decades as a result of this technology.

1. Economic crimes and identity theft
2. Intellectual property theft and piracy
3. Cyber-stalking and online harassment
4. Hacktivism and online extremism
5. Cyber-terror and cyber-warfare
6. Computer hackers and hacking

Economic Crimes and Identity Theft

The ability to buy and sell goods at any time or location is revolutionary for both commerce and cybercrime. Email is a critical resource for attackers who target victims for mass exploitation. They may stress a time-sensitive problem to compel their victims to impart personally identifiable information to use for fraud and/or impersonation.

The emergence of "carding markets" (credit card data on the Darknet) has increasingly fueled the demand for this information. Interestingly, many victims of these crimes report the fraud to banks to recover account funds but never report the crime to law enforcement agencies.

Intellectual Property Theft and Piracy

Ideas become *intellectual property* when they are physical and tangible works that may be accessed and viewed by others. Low-income countries are the leaders in pirating software and their changing tactics and behaviors make it difficult to prevent these types of crimes.

Utilizing bit-torrent links to a multitude of "seeder servers" (servers containing pirated content) allow concurrent uploads and downloads to be executed quickly and reduces the risk and likelihood of detection.

Listed below are examples of intellectual property that are common targets of digital piracy.

1. Copyrights
2. Trademarks
3. Patents
4. Trade secrets (strategies and processes)

Cyber-Stalking and Online Harassment

The open nature in which people can lead their lives is unparalleled—except by the access they provide to their own personal information. This crime involves constant use of email and text and has the potential to cause real-world harm. Victims of these crimes report feeling anxiety, depression, and a feeling of powerlessness.

Victims may reduce exposure by changing their behaviors, contributing to elevated levels of absenteeism from work and social gatherings. In both cyber-stalking and online harassment, the offender does not need to know the person being victimized.

Hacktivism and Online Extremism

A carefully targeted attack against any critical infrastructure asset can cause serious harm to the functionality and security of the network. "Hacktivists" utilize technology to promote an activist agenda. The primary differences between hacktivism and online extremism are the motives of the actor and the number of people harmed.

Online extremists tend to target larger groups of people with the intention of impacting innocent people. In addition to causing panic and fear in the general population, these actions can cause mass collateral damage on a physical level that rivals traditional terrorist attacks.

Hacktivism in Action: *Anonymous*

Anonymous is a decentralized and international hacktivist group known for its cyber-attacks against governments, institutions, agencies, and corporations. The collective, associated with online hacktivism, originated in 2003 and is comprised of members known as "*Anons*."

The group operates domestically and internationally motivated by several issues and ideologies. Because no definitive leadership exists, attribution for group actions is difficult. The collective is also unique as they also engage in direct physical actions offline.

Cyber-Terror and Cyber-Warfare

The ability to communicate with others of diverse backgrounds ensures that individuals can be steadily introduced to a movements' principles. Although it is well-established that cyber-warfare is a very real threat, there is no agreed upon definition of it. Social media plays a key role in cyber-terrorism; Twitter is a key platform for the radicalization of individuals.

Groups such as "Al Qaeda" use websites to publish the magazine *Inspire*, a publication dedicated to providing detailed tactics for conducting terrorist activities.

"eJihad" online forums such as *al-Farouq* and *al-Jinan* provide their visitors with the ability to obtain free exploit tools to compromise systems belonging to entities with opposing ideologies. In the United States, supremacist groups such as the *Ku Klux Klan* and *Aryan Nations* are placed into this category.

Computer Hackers and Hacking

A "hack" involves the modification of technology in order to allow it to be used in unintended ways for both legal and illegal purposes. Many hackers use non-technical strategies such as social engineering, realizing that humans will always be the weak link in the security chain.

The technology used is ethically neutral; the same tools used by hackers are the same as those used to attempt to prevent them from hacking.

They tend to commit a significant amount of time to studying their craft and form online relationships to both assist in their studies and validate their behaviors. Hackers typically fall into one of three categories: *white hat* (good), *gray hat* (good and bad), and *black hat* (criminals). When black hat hackers engage in the exploit of a network, they follow a pattern referred to as the *Attack Life Cycle*.

Script Kiddies (aka: "Skiddies" or "Skids")

A *Script Kiddie* is a person who is not technologically sophisticated, who randomly seeks out a specific weakness to gain access to a network or system. They tend to be less experienced with exploits and possess less developed hacking skills, relying instead on existing and proven techniques and scripts. They tend to be motivated by basic and personal reasons – such as fun, chaos, attention, and revenge – and have little regard for the potential consequences of their actions.

Script Kiddies: An Underestimated Threat

A Script Kiddie can easily find vulnerabilities to access private systems using programs intended for computer security and attack networks and websites. Social engineering is a tactic that is used extensively – phishing, vishing, and fake websites – and their actions can cause extreme damage and/or disruption.

While skilled hackers rely on planned exploits, members of this category tend to be unpredictable. A real threat is posed in that they may unwittingly aid serious criminals, exposing themselves so that the sophisticated actors they serve can remain hidden. Although frequently underestimated, they are unquestionably cyber criminals.

Legal Challenges in Digital Intelligence Collection

Evolving privacy and data protection regulations across diverse geographical locations may add to the complexity of gathering the digital evidence required for effective threat intelligence operations. The following sections will focus on five prominent challenges experienced by analysts on a regular basis.

1. The Electronic Communications Privacy Act
2. The Fourth Amendment
3. The Fifth Amendment
4. International Jurisdictions
5. *Nulla Poena Sine Lege*

The Electronic Communications Privacy Act (ECPA)

The *Electronic Communications Privacy Act* (ECPA) was passed in 1986 to expand and revise existing wiretapping and electronic eavesdropping provisions. This legislation was adopted to serve three primary purposes:

1. Add amendments to the *Wiretap Act*
2. Add updates to the *Stored Communications Act*
3. Establish the *Pen Register Act*

The goals of the ECPA were to bar electronic eavesdropping when not authorized by an official warrant, a prevalent activity in many non-legislated digital environments. It banned disclosure of data that was illegally obtained—typically via the unlawful access of data at rest.

It also prohibited the use of pen register and trace devices commonly used in warrantless searches to develop probable cause during investigations.

The Fourth Amendment

The *Fourth Amendment* limits the legal authority of government—or agents of the government—to search and seize evidence without a warrant. While this may be widely known, its application to digital storage devices is complicated. The Fourth Amendment only applies to government agencies but may present difficulties for organizations attempting to submit evidence in court for legal actions.

The Fourth Amendment:

> *The right of the people to be secure in their persons, houses, papers, and effects, against unreasonable searches and seizures, shall not be violated, and no Warrants shall issue, but upon probable cause, supported by Oath or affirmation, and particularly describing the place to be searched, and the persons or things to be seized.*

Courts have ruled that all individuals have a reasonable expectation of privacy that includes their body, home, and possessions. Courts have also ruled modern phones hold the "privacies of life" and must be treated like computers.

Exceptions to these rules include exigent circumstances in the event of an emergency, and the *Shared Authority Principle* which applies to authority that is shared by virtue of a relationship (such as an agreement between employer and employee). In all cases, consent is the most powerful and relevant exception to this rule.

The Fifth Amendment

The *Fifth Amendment* protects people from self-incrimination such as giving a statement that might expose oneself to prosecution. It also establishes that silence is not an indication of guilt and that compelling a person to relinquish this right to silence is a violation of the rule.

The Fifth Amendment:

> *No person shall be held to answer for a capital, or otherwise infamous crime, unless on a presentment or indictment of a Grand Jury, except in cases arising in the land or naval forces, or in the Militia, when in actual service in time of War or public danger; nor shall any person be subject for the same offence to be twice put in jeopardy of life or limb; nor shall be compelled in any criminal case to be a witness against himself, nor be deprived of life, liberty, or property, without due process of law; nor shall private property be taken for public use, without just compensation.*

This presents some unique and challenging circumstances when dealing with digital devices. A few of these challenges include access codes, device encryption, and biometrics. There are no *Key Disclosure* laws in the United States, leaving interpretation of these issues up to courts on a case-by-case basis. As with the Fourth Amendment, consent is the most powerful and relevant exception to this rule.

Nulla Poena Sine Lege

Nulla poena sine lege is a universal legal principle, requiring that one cannot be punished for doing something that is not prohibited by law. This basic requirement for the rule of law is challenging when laws cannot keep pace with technology and when sovereign jurisdictions cannot agree on which actions constitute a crime.

There are four elements of this legal philosophy that impact the ability to conduct intelligence operations and prosecute crimes in the digital environment:

1. *Nulla Poena Sine Lege* ***Praevia***: The law must exist prior to the crime and cannot be applied "after the fact."

2. *Nulla Poena Sine Lege* ***Scripta***: The law must be written and cannot be based on opinion or customary law.

3. *Nulla Poena Sine Lege* ***Certa***: The law must be well-defined such that any citizen would know the action is a crime.

4. *Nulla Poena Sine Lege* ***Stricta***: The law must be exact and prohibits the application of analogous applications.

CHAPTER 03

Adversaries in the Modern Threat Landscape

Knowledge Assessment Questions

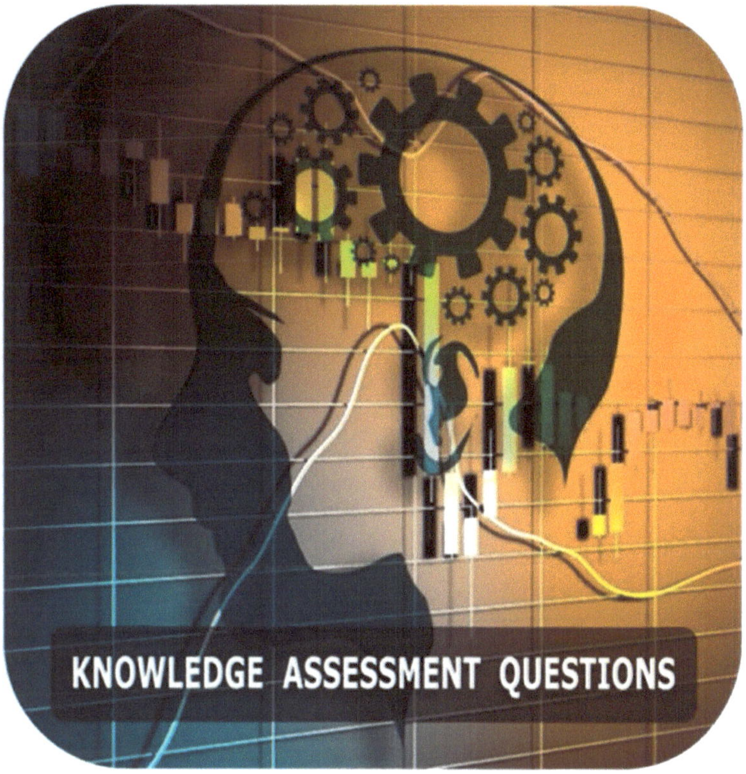

NOTICE TO STUDENTS:

The following knowledge assessment questions are presented in true / false, multiple choice, and fill-in-the-blank formats. The correct answers are provided in an Answer Key at the end of Chapter 14. These questions may or may not be presented on the quizzes and/or certification exam given by the instructor of this course.

Knowledge Assessment Questions

1) What choice below is a result of passing the *Electronic Communications Privacy Act* (ECPA) in 1986, legislation intended to expand and revise wiretapping and electronic eavesdropping provisions?

A. Created the *Wiretap Act*

B. Created the *Stored Communications Act*

C. Created the *Pen Register Act*

D. Expanded Electronic Eavesdropping

2) What type of hacker matches the following definition: "*A person who is not technologically sophisticated, who randomly seeks out a specific weakness to gain access to a network or system*"?

A. White Hat

B. Script Kiddie

C. Gray Hat

D. Ethical

3) What hacktivist group matches the following definition: "*A decentralized and international hacktivist group known for its cyber-attacks against governments, institutions, agencies, and corporations*"?

A. Anonymous

B. Elfin Team

C. Ocean Lotus

D. Fancy Bear

4) Ideas become __________ when they are physical and tangible works that may be accessed and viewed by others.

A. Profitable

B. Commercially Available

C. Time Sensitive

D. Intellectual Property

5) A(n) __________ is a threat actor, typically a nation-state or state-sponsored group, attempting to gain unauthorized access to a network.

A. Hacktivist Group

B. Script Kiddie

C. Advanced Persistent Threat

D. Nonstate Hacker

True / False

6) The ability to communicate with others of diverse backgrounds ensures that individuals can be steadily introduced to a movement's principles.

A. True

B. False

7) An abstraction involves the modification of technology in order to allow it to be used in unintended ways for both legal and illegal purposes.

A. True

B. False

8) Threat intelligence can be derived from external sources such as open-source platforms, or communications between threat information sharing groups.

A. True

B. False

9) The Fourth Amendment protects people from self-incrimination, meaning giving a statement that might implicate oneself in a crime.

A. True

B. False

10) Evolving privacy and data protection regulations across geographies may add to the streamlining of gathering actionable intelligence and accessing sources.

A. True

B. False

CHAPTER 04

Enhancing Analytical and Logical Reasoning Skills

Lecture Summary

KEY KNOWLEDGE POINTS

Overview of Basic Thinking Abilities

Critical Thinking: The Elements of Thought

The 8 Intellectual Standards of Analysis

Avoiding Analytical Pitfalls

An Overview of Biases

An Overview of Logical Fallacies

Overview of Basic Thinking Abilities

When communicating the results of analysis, the analyst must ensure that the customer understands the conclusions and how they were reached. To accomplish this successfully – whether discussing basic abilities of complex skills – relates directly to the analyst's ability to think. This requires continual improvement in the quality of thinking. In the discipline of intelligence analysis, the three basic thinking abilities listed below are required.

1. Information Ordering
2. Pattern Recognition
3. Logical Reasoning

Information Ordering

Information ordering is the ability to follow predefined rules or sets of rules to arrange the data in a meaningful order that enhances the intelligence process. This allows analysts to arrange information, often with the assistance of technology, to perform analysis, synthesis, and gain a higher understanding of the issue or problem. The process helps the analyst make conclusions which are then disseminated as an intelligence product. The risk of information ordering is that it is inherently limiting; the analyst may fixate on a specific order and not recognize alternative explanations.

Pattern Recognition

Analysts detect and impose *patterns* on apparently random entities and events in order to understand them, often doing this without being aware of it. Patterns are imposed to detect relationships, infer potential actions in the future, conceptualize a degree of order from chaos, and separate relevant facts from irrelevant information. Look at the graphic below and determine if any patterns exist that may indicate an employee is falsifying employment information.

EMPLOYEE "**A**"		EMPLOYEE "**B**"	
Employer Information		**Employer Information**	
Employer:	Geek Tech Computer Repair	Employer:	Savannah Bytes Programming
Address:	438 Mooreland Avenue NE	Address:	124 Stevens Court SW
City, State, Zip Code:	Marietta, Georgia 30062	City, State, Zip Code:	Savannah, Georgia 30124
Telephone:	(770) 794-9390	Telephone:	(912) 487-5050
Applicant Information		**Applicant Information**	
First Name:	Nicholas	First Name:	Michael
Last Name:	Ferron	Last Name:	Guifanno
Address:	829 Jefferson Street NW	Address:	1 Riverside Drive, Apt. 24a
City, State, Zip Code:	Atlanta, Georgia 30308	City, State, Zip Code:	Rincon, Georgia 31247
Telephone:	(404) 506-1923	Telephone:	(912) 493-3446
Date of Birth:	11/15/1977	Date of Birth:	01/24/1978
Social (SSN):	157-43-1560	Social (SSN):	250-12-4693
Driver License:	(GA) 047829025	Driver License:	(GA) 030612477

The risk associated with seeking patterns is that it has the potential to introduce bias. The analyst may impose culturally defined patterns or not be able to recognize inherent patterns. An analyst that unintentionally imposes patterns that do not exist may overlook the pattern inherent in the information of employee "B" shown below. That pattern is the recurring number sequence 1, 2, and 4.

EMPLOYEE "**A**"		EMPLOYEE "**B**"	
Employer Information		**Employer Information**	
Employer:	Geek Tech Computer Repair	Employer:	Savannah Bytes Programming
Address:	438 Mooreland Avenue NE	Address:	**124** Stevens Court SW
City, State, Zip Code:	Marietta, Georgia 30062	City, State, Zip Code:	Savannah, Georgia 30**124**
Telephone:	(770) 794-9390	Telephone:	(9**12**) **4**87-5050
Applicant Information		**Applicant Information**	
First Name:	Nicholas	First Name:	Michael
Last Name:	Ferron	Last Name:	Guifanno
Address:	829 Jefferson Street NW	Address:	**1** Riverside Drive, Apt. **24**a
City, State, Zip Code:	Atlanta, Georgia 30308	City, State, Zip Code:	Rincon, Georgia 3**124**7
Telephone:	(404) 506-1923	Telephone:	(9**12**) **4**93-3446
Date of Birth:	11/15/1977	Date of Birth:	0**1**/**24**/1978
Social (SSN):	157-43-1560	Social (SSN):	250-**12**-**4**693
Driver License:	(GA) 047829025	Driver License:	(GA) 0306**124**77

Logical Reasoning

The ability to *reason* is what permits analysts to process information and formulate explanations, to assign meaning to observed actions and events. It allows raw information to be converted into an intelligence product. The four primary types of reasoning which guide analysts are listed below.

1. Inductive Reasoning
2. Analogical Reasoning
3. Deductive Reasoning
4. Abductive Reasoning

Inductive Reasoning

Inductive reasoning is a process of discovery in which an analyst establishes a relationship between events under observation or study. It normally precedes deduction and is the most common type of analytical reasoning. This type of reason is dependent upon accurate observation and can lead to now sources of knowledge.

The three requirements needed for inductive reasoning to be successful are *objectivity*, the *elimination of prejudice*, and the *elimination of preconceptions* on the part of the analyst.

Analogical Reasoning

Analogical reasoning is a method of processing information that compares the similarities between new concepts and previously understood concepts. In this type of reasoning, similarities are used to gain understanding. If there are substantial similarities between previous known situations and current unknown situations, there are likely to be further similarities that can be discovered. Analysts can use facts known about both situations to infer new information that may not have been initially obvious.

Deductive Reasoning

Deductive reasoning is a logical process in which a conclusion is based on the concordance of multiple premises that are generally assumed to be true. Analysts begin with a set of rules – an empirical methodology – which are then used to interpret information. The final argument is sound if the original premise is true.

The risk of this type of reasoning is that it may lead to an incorrect conclusion. If the analyst fails to properly interpret a series of events preceding an action, mistakes can be made confusing cause and effect. For this reason, deductive reasoning is not effective when forecasting human behavior.

Abductive Reasoning

Abductive reasoning describes the logical thought process that accompanies the insight, intuition, or experience of the analyst conducting the research. However, this type of reasoning is prone to error if the analyst does not impose objectivity. As a result, the available information is not consistent with the expected outcome, and a reason for this discrepancy must be identified. A new hypothesis can be generated by disassociating from the original premise, re-ordering the available information, then re-engaging in the analytical process. The new explanation that is revealed must then be tested for validity.

Critical Thinking: The 8 Elements of Thought

Analysts often tend to believe their thoughts are accurate just because they are theirs; therefore, the thoughts of those that disagree with them are believed to be inaccurate. The eight *elements of thought* listed below are used in all analytical processes and occur naturally in all rational human beings.

1. Purpose and Goals
2. The Question at Issue
3. Available Information
4. Interpretation and Inference
5. Accepted Concepts
6. Assumptions and Opinions
7. Implications and Consequences
8. Individual Point of View

The 8 Intellectual Standards of Analysis

While the *elements of thought* provide a framework for **analyzing** thinking, the *standards of thought* provide criteria to assess the **quality** of thinking. They provide the analyst with a solid foundation for *critical thinking*. Critical thinking is a deliberate choice that requires focus, self-discipline, and a commitment to continuously improve skills. When successfully employed, the eight standards of thought listed below ensure analysts are effective and efficient when approaching either the most complicated or ambiguous tasks.

1. *Fair-Mindedness*: This entails an awareness that humans tend to prejudge the views of others, placing them into favorable (agreement) or unfavorable (disagreement) categories.

2. *Intellectual Humility*: This is the knowledge of ignorance and the analyst's sensitivity to what they know and do not know.

3. *Intellectual Courage*: This is the disposition of the analyst to question beliefs they feel strongly about and a willing to express views even when they are unpopular.

4. *Intellectual Empathy*: This is awareness of the need to actively entertain different and opposing views, especially those with which the analyst strongly disagrees.

5. *Intellectual Integrity*: This consists of the analyst holding themselves to the same intellectual standards they expect others to honor (disallowing double standards).

6. *Intellectual Perseverance*: This is the disposition of the analyst to work their way through intellectual complexities despite the frustration inherent to the task.

7. *Confidence in Reason*: This requires using standards of reasonability as the fundamental criteria by which to judge whether to accept or reject any belief or position.

8. *Intellectual Autonomy*: This requires the analyst to think through issues or problems using their own thinking rather than uncritically accepting the viewpoints of others.

Avoiding Analytical Pitfalls

Elements of thought, intellectual standards, and intellectual traits aid analysts in recognizing pitfalls in their own analysis and the analysis performed by others. Analysts and the analysis process are subject to numerous influences. In analysts, these influences serve to shape and mold their view of the world and shape their ability to reason. These influences are referred to as *analytical pitfalls*. The two categories of pitfalls discussed in this course are *biases* and *logical fallacies*.

Biases

A *bias* is a disproportionate weight in favor of or against an idea or thing, usually in a way that is closed-minded, prejudicial, or unsupported by fact. Biases generally have a detrimental impact on intelligence analysis because they obscure the true nature of the information (both received and disseminated). The four categories of bias listed below are the types most commonly observed in analysts.

1. *Cultural Bias*: This category of bias is often referred to as "Mirror Imaging."

2. *Organizational Bias*: This category of bias includes "Group Think" and "Best Case" reasoning.

3. *Personal Bias*: This category of bias includes "Assimilation Bias," "Conformation Bias," "Self-Verification Bias," and "Mental Anchoring Bias."

4. Cognitive Bias: This category of bias includes "Vividness," "Oversensitivity to Consistency," "Dependency on Memory," and "Acceptance of New Intelligence."

Logical Fallacies of Relevance

Although usually committed unintentionally, these fallacies are sometimes used deliberately to persuade, convince, or deceive those unaware that they exist. An analyst must be able to recognize logical fallacies so a false line of reasoning will not distract them and lead to poor, unsupported conclusions. The eight categories of relevance fallacies listed below are the types most observed in analysts, especially when engaged in deliberations in group settings.

1. *Appeal to Force*: This category of fallacy presents itself as a last resort in the form of intimidation when logic fails to persuade.

2. *Genetic Fallacy*: This category of fallacy assumes that an opposing idea is wrong based on its origin (cultural, regional, racial, ethnic, or religious).

3. *Argument to the Man*: This category of fallacy seeks to discredit the source of the opposing viewpoint to win the argument.

4. *Fallacy of Hasty Generalizations*: This category of fallacy attempts to reach a conclusion based on the oversimplification of facts associated with complex issues.

5. *Fallacy of Division*: This category of fallacy assumes the whole represents all of its parts, thereby oversimplifying the complexity and diversity of the elements and sub-elements of the whole.

6. *Post Hoc Ergo Propter Hoc*: This category of fallacy – Latin meaning "after this, therefore because of this" – states "*since event Y followed event X, event Y must have been caused by event X*." This is the root cause of distorting reasons for cause and effect relating to issues and problems.

7. *Fallacy of False Dilemma*: This category of fallacy – also referred to as "Black and White Thinking" – presents a false choice based on limited options that do not take into consideration the complexities associated with issues or problems (if you're not with me, you're against me).

8. *Fallacy of Special Pleading*: This category of fallacy is based on the establishment of a universal principal but requires the universal principal to be overlooked as nonapplicable to the issue or problem being debated. In such cases, the existence of a universal truth that cannot be debated which is the foundation of a logical argument – but then gets dismissed by the same individual to win an argument – is not logically consistent.

Logical Fallacies of Omission

Fallacies of omission occur when an analyst leaves out necessary material (either intentionally or unintentionally) in a conclusion or inference. The three categories of omission fallacies listed below are the types most observed in analysts, especially when engaged in deliberations in group settings.

1. *Oversimplification*: This category of fallacy uses generalities as the logical foundation for an argument that fail to take into consideration and account for the complexity of the issue or problem. This is similar to *Fallacy of Hasty Generalizations*. However, instead of oversimplifying facts to reduce their complexity, complex facts are ignored entirely so as not to have to be considered in the final analysis.

2. *Fallacy of Composition*: This category of fallacy assumes that the features and characteristics of elements of the whole are representative of the whole itself. It presents itself as the opposite of the *Fallacy of Division* discussed earlier in this section.

3. *Misusing Analogies*: This category of fallacy occurs when analysts generalize indiscriminately from an analogy to a real-world event or employ counter-analogies during argumentation which assume dissimilar events have similar characteristics when none exist.

CHAPTER 04

Enhancing Analytical and Logical Reasoning Skills

Knowledge Assessment Questions

NOTICE TO STUDENTS:

The following knowledge assessment questions are presented in true / false, multiple choice, and fill-in-the-blank formats. The correct answers are provided in an Answer Key at the end of Chapter 14. These questions may or may not be presented on the quizzes and/or certification exam given by the instructor of this course.

Knowledge Assessment Questions

1) What choice below is one of the three basic thinking abilities required by intelligence analysts to be successful in the performance of their duties?

A. Logical Recognition

B. Dependency on Memory

C. Premature Closure

D. Information Ordering

2) __________ reasoning is a process of discovery in which an analyst establishes a relationship between events under observation or study.

A. Inductive

B. Abductive

C. Deductive

D. Analogical

3) A(n) __________ is a disproportionate weight in favor of or against an idea or thing, usually in a way that is closed-minded, prejudicial, or unsupported by fact.

A. Intellectual Standard

B. Bias

C. Element of Thought

D. Logical Fallacy

4) __________ reasoning is a logical process in which a conclusion is based on the concordance of multiple premises that are generally assumed to be true.

A. Analogical

B. Inductive

C. Deductive

D. Abductive

5) When analysts dissect their thinking and examine its parts, they use standards of quality referred to as __________ or *standards of thought*.

A. Elements of Thought

B. Critical Thinking

C. Logical Reasoning

D. Intellectual Standards

True / False

6) *Fallacies of omission* occur when an analyst leaves out necessary material (either intentionally or unintentionally) in a conclusion or inference.

A. True

B. False

7) While the *elements of thought* provide a framework for qualifying thinking, the standards of thought provide criteria to assess the analysis of thinking.

A. True

B. False

8) When communicating the results of analysis, the analyst must ensure that the collection manager understands the conclusions and how they were reached.

A. True

B. False

9) Analysts detect and impose patterns on apparently random entities and events in order to understand them, often doing this without being aware of it.

A. True

B. False

10) Abductive reasoning is a method of processing information that compares the similarities between new concepts and previously understood concepts.

A. True

B. False

CHAPTER 05

Managing the Information Collection Process

Lecture Summary

KEY KNOWLEDGE POINTS

The Role of the Collection Manager

Collection Requirements Management Process

Conducting a Gap Analysis

Analyst and Collector Collaboration

Overview of the Collection Strategy

Developing a Collection Plan

Chapter 05: *Managing the Information Collection Process*

The Role of the Collection Manager

In the field of intelligence, the facilitator between the producers of intelligence (analysts) and the operators (collectors) is the Collection Manager. Collection Manager plays a critical role in the functional areas of responsibility listed below.

1. Communicates the analyst's needs to the collectors
2. Generates collection requirements
3. Responds to identified intelligence gaps
4. Understands the collector's capabilities
5. Understands the analyst's needs
6. Selects the appropriate discipline for the collection requirement

The Collection Manager ensures and maintains an efficient two-way line of communication always exists between the analysts and the collectors. Facilitating feedback to the collectors ensures the collected information meets the operation's needs. Additional information may be required, and analysts may have questions about the information already collected. This continuous feedback helps the collectors improve their process.

The Collection Manager operates across all echelons, all security and network domains, and across all computing environments to aid intelligence personnel. Listed below are several responsibilities performed by the Collection Manager to ensure this aid is provided to the appropriate personnel.

1. Data mining and research
2. Collaboration between teams and units in real time
3. Conducting link and pattern analysis
4. Developing threat and warning templates
5. Insulating the team members from external forces
6. Maintaining intelligence databases and files
7. Disseminating information amongst appropriate teams

Analysis Scenario: The Air Force Generals Disagree

This analysis scenario is designed to have students think through possible solutions using logical reasoning skills. The logical reasoning skills used to evaluate this scenario are equally as important as the answer. The directions for this scenario are listed below.

1. **Situation**: An additional 500 pounds of armoring has been authorized for a surveillance aircraft. One General wants to armor the avionics (electronic systems) in the nose of the aircraft, the other General wants to armor the fuselage (body) of the aircraft. Only **one** location can be chosen.

Your team has been sent to a base to analyze the situation. Every day you observe **10** surveillance aircraft leave the base on missions. **80%** of the aircraft return to the base after completing the mission. It is reported by returning pilots those that did not return were **shot down** during the mission.

You notice that **100%** of the returning aircraft have a significant number of bullet holes in the fuselage and **no** damage to the avionics. This continues for **every** mission. You document the results of **10 missions** and are asked to advise the Generals where the armor must be placed to protect the planes.

2. **IR (Information Requirement) Tasking Order**: Based on your analysis, where would you recommend placing the additional armor on the aircraft? The avionics, or the fuselage

3. **Rules of Engagement**: No actions may be taken to alter the environment or any of the components presented in this scenario. Nothing may be added, removed, or manipulated by participants in this analysis. All facts have been presented; there is nothing "hidden." However, there may be questions needed for clarification. If so, ask your course instructor.

Final Analysis: The Air Force Generals Disagree

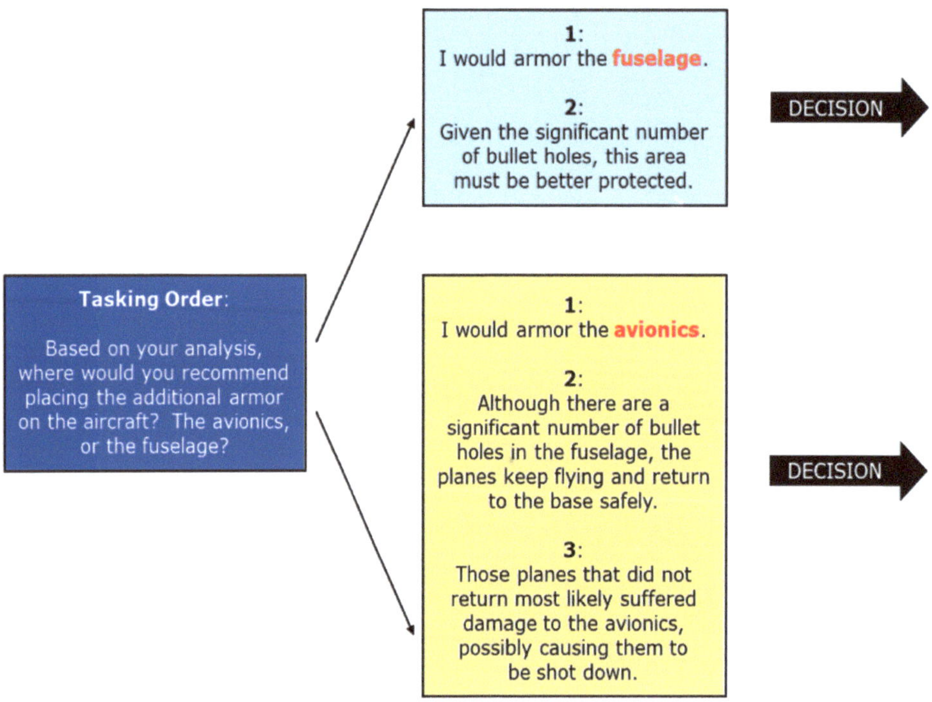

Collection Requirements Management Process

1. Develop the Requirements
2. Develop the Collection Plan
3. Task or Request the Collection
4. Manage the Dissemination

5. Evaluate the Reporting

6. Update the Collection Planning

Step 1: Develop the Requirements

As the collection process is used to satisfy requirements to fill gaps, it is simultaneously used to generate new requirements or reprioritize existing ones. After participating in testing or simulation, the requirements (intelligence gaps) are analyzed. They are then recorded, validated, consolidated, and prioritized. The Collection Manager can then develop *Specific Information Requests* (SIRs).

Step 2: Develop the Collection Plan

The goal is to develop a collection and tasking plan that produces the intelligence required to effectively answer the customer's information requirements. The Collection Manager will evaluate the available resources and develop a collection strategy. This process includes selecting the appropriate resources for the task and synchronizing the collection effort to the requirements. Once completed, the Collection Manager will develop *Specific Orders and Requests* (SORs) and prioritize them for the collection assets.

Step 3: Task or Request the Collection

The benefit of communicating *collection intent* is the holistic perspective it provides internal and external collectors isolated from the planning process. As part of this communication, the Collection Manager will determine the tasking mechanism, determine the request mechanism, and execute the implementation of both. It is as this stage the team can begin the collection and exploitation of the required intelligence.

Step 4: Manage the Dissemination

The Collection Manager will arrange for the direct dissemination of the collected intelligence. This requires determining the levels of perishability and the amount of information to disseminate. Once the media for dissemination has been identified the collected information can be disseminated to the appropriate recipients. Whenever possible, write in the *Specific Orders and Requests* (SORs) the requirement for direct dissemination of intelligence to the customer of the product.

Step 5: Evaluate the Reporting

All *Specific Orders and Requests* (SORs) must be fully satisfied in a timely manner, keeping the intelligence system fully synchronized. This requires continuous monitoring of the process and maintaining the synchronization. Reports are correlated to the requirements and then screened for relevance. This then allows the Collection Manager to provide feedback to both the collectors and the exploiters of the intelligence.

Step 6: Update the Collection Planning

Specific Orders and Requests (SORs) are updated to keep intelligence synchronized with the operation, and collectors work only on unsatisfied requirements. Satisfied requirements can be eliminated from the tasking and assets can be redirected to those requirements still unsatisfied. The Collection Manager will then cue assets to potential collection opportunities, maintain continuous synchronization of the collection, and add new collection requirements as needed.

Conducting a Gap Analysis

Identified gaps need to be made explicit, and the intelligence collection effort must fill in the missing pieces of the model to answer the customer's questions. This is accomplished by using a *gap analysis process*. The Collection Manager can prioritize identified gaps based on importance or size. The gaps are then classified by their nature – collection, processing, or analysis – and then sorted again based on timing (short-term or long-term).

Overview of the Collection Strategy

When true collection gaps have been defined, it is time to plan for collection to fill them and offer guidance to the sources of collection. The gaps about the target must be identified and the nature of the gap must be understood. Not all gaps are collection gaps.

1. Data about the target can be inadequate or nonexistent.
2. Redundant collection efforts waste valuable time.
3. Unneeded efforts do not solve intelligence problems.
4. Excess data compounds the problem and the workload of all involved.

Analyst and Collector Collaboration

The secret to any successful intelligence collection strategy is a sustainable and close relationship between analysts and collectors on the team. Collectors tend to have a good understanding of the issue which allows them to provide effective responses to requests. Analysts tend to lack the knowledge of the collectors and need to dedicate time to understanding collection capabilities. Individual collectors have different operational cultures and focuses. If analysts can understand the preferences of these collectors (based on culture and issues) the relationship between the two can be strengthened.

Basic Collection Strategy Development

Most collection strategies are developed because of past experiences; use of existing methods saves time, effort, and organizational resources. The Collection Manager should look at successes from previous efforts. Collection strategies against similar targets are identified and the results of similar strategies are measured. However, the effort cannot rely solely on previous strategies as predictable collection efforts can be defeated. Innovative methods of executing previously successful strategies should be developed, and the timing of the collection for the most desirable content should be established.

Developing a Collection Plan

"Garbage in, garbage out" is appropriate in understanding how effective the intelligence team is in responding to the customer's requirements. Collectors and analysts cannot guess the customer's needs. To avoid this issue, planning and direction is a foundational requirement. The tasks listed below must be satisfied by customers of intelligence products.

1. The customer must **articulate** their specific needs.
2. The customer must **validate** all intelligence requirements.
3. The customer must **address** specific questions regarding requirements.
4. The customer must **identify** all known intelligence gaps that currently exist.

The Collection Requirements Process

Intelligence collection requirements fall within two categories: tasking *organic resources* belonging to the organization or requesting use of *external resources*.

1. **Organic Resources**: Intelligence Analysts, SOC Cybersecurity, Risk Management, and Corporate Security.
2. **External Resources**: Upper Echelon Capabilities, Law Enforcement, Government, and Academia.

[THIS PAGE IS LEFT INTENTIONALLY BLANK.]

CHAPTER 05

Managing the Information Collection Process

Knowledge Assessment Questions

NOTICE TO STUDENTS:

The following knowledge assessment questions are presented in true / false, multiple choice, and fill-in-the-blank formats. The correct answers are provided in an Answer Key at the end of Chapter 14. These questions may or may not be presented on the quizzes and/or certification exam given by the instructor of this course.

Knowledge Assessment Questions

1) In the field of intelligence, the facilitator between the producers of intelligence (analysts) and the operators (collectors) is the ___________.

A. Collection Manager

B. Fusion Analyst

C. Team Leader

D. Intelligence Customer

2) The benefit of communicating "___________ intent" is the holistic perspective it provides internal and external collectors isolated from the planning process.

A. Processing

B. Collection

C. Exploitation

D. Dissemination

3) When true collection ___________ have been defined, it is time to plan for collection to fill them and offer guidance to the sources of collection.

A. Plans

B. Methods

C. Gaps

D. Sources

4) Collection managers cannot rely solely on previous strategies because predictable collection methods can be ___________.

A. Repeated

B. Modelled

C. Implemented

D. Defeated

5) When collection managers are developing the initial collection plan, it is imperative that customers ___________ their specific needs.

A. Articulate

B. Question

C. Refine

D. Support

True / False

6) The Collection Manager operates across all echelons, all security and network domains, and across all computing environments to aid intelligence personnel.

A. True

B. False

7) Identified sources need to be made explicit, and the intelligence collection effort must fill in the missing pieces of the model to answer the customer's questions.

A. True

B. False

8) Intelligence collection requirements fall within two categories: tasking external resources belonging to the organization or requesting use of future resources.

A. True

B. False

9) Most collection strategies are developed as a result of past experiences; use of existing methods saves time, effort, and organizational resources.

A. True

B. False

10) Whenever possible, write into the *Specific Information Requirements* the requirement for direct dissemination of intelligence to the customer of the product.

A. True

B. False

CHAPTER 06

Determining the Root Cause of Intelligence Issues

Lecture Summary

KEY KNOWLEDGE POINTS

The Intelligence Issue

Preliminary Research Questions

Identifying Issue Categories

The PMESII / ASCOPE Matrix

The Key Assumptions Check

Issue Decomposition

Chapter 06: *Determining the Root Cause of Intelligence Issues*

The Intelligence Issue

All intelligence analysis efforts begin with an *issue* or problem definition which aligns with the customer's interest about the target. The issue must be understood in detail by all those involved in the process. Determining the need for the request allows for the making of proper decisions the results will support. All successful intelligence operations must begin with an accurate issue definition. This also allows members of the intelligence team to discern important information from irrelevant data.

Veteran vs. Novice Analysts

The initial guidance customers give analysts about an issue is almost always incomplete and may be unintentionally misleading. Veteran analysts know this and approach the process differently than novice analysts. Listed below are several actions taken by veteran analysts that help to make their efforts more efficient and effective.

1. Spend more time initially considering the issue or problem.
2. Make a conscious effort to actively avoid the *framing effect*.
3. Use knowledge of previous cases to provide clarity.
4. Use the proper context for creating mental models.
5. Recognize when information is lacking and know how to react.
6. Focus the majority of time on the problem definition phase.

Preliminary Research Questions

In the discipline of intelligence analysis, certain guidelines (based on scientific research) are generally followed before starting a new project which are based on the five fundamental questions listed below.

1. Who is the **customer**?
2. What is the **purpose**?
3. What are the **real** questions?
4. When are the **answers** required?
5. What intelligence product will be most **effective**?

Question 1: Who is the Customer?

After establishing parameters and obtaining guidance with these questions, analysts will have a starting point for defining the issue or problem statement. The intelligence customers must be identified and their specific needs understood. This process may involve several intermediaries, a reality that may pose challenges. Needs tend to become distorted as they move through long channels. To overcome this challenge, analysts should seek direct interaction with the customer if possible and consider second-order recipients of the intelligence product as well.

Question 2: What is the Purpose?

On large and complex intelligence projects, the *parameter statement* will itself become a formal product to be disseminated to stakeholders. Intelligence efforts tend to have one primary purpose that should be clear to everyone when the effort begins and in the expected result. The customer's involvement in the process will help to facilitate clarity.

Question 3: What are the Real Questions?

The issue definition product helps to explain the real questions to be answered and all the related issues which need to be addressed. Analysts must obtain all relevant background information underlying the customer's problem. They should also understand the problem's real impact on policy and operational decisions. Vaguely worded requests can be misleading and result in an intelligence product that does not meet the customer's needs.

Question 4: When are the Answers Required?

Once the issue definition product is complete, analysts will be able to focus on answering the questions that are relevant and meet customer expectations. All intelligence operations have time requirements, so analysts must know when the product needs to be delivered. Many products are delivered too late – customers receive reports after decisions have been made – which tends to be a result of the customer being isolated from the process. The target-centric models speed up the process by having the customer integrated into the overall effort.

Question 5: What Product will be most Effective?

A well-defined statement of the problem (the result of these answers) still requires two additional steps: *issue definition* and *issue decomposition*. The results are then disseminated to the customer. Written reports (digital) are the standard format as they endure over time and are easily distributed to multiple customers. Verbal briefings are another option – if the customer prefers this method. Briefings allow customer interaction, feedback, and ensure the intended recipient receives the message in a timely manner.

The Issue Definition

When analysts receive the initial tasking (from a customer of self-initiated) time pressures may tempt them to give the issue definition minimal consideration. Improperly defined issues have the potential to void the analysis effort. Customers rarely ask the right questions. They do not ask for what they *need* to know, but rather what they *want* to know, or what they *think* they need to know. Analysts must be able to under the difference and guide the customer accordingly. Understanding gaps in the customer's knowledge will help to make this possible.

Identifying Issue Categories

Most intelligence needs, reflected in the questions customers will likely ask about a target, fall into one of the five general *issue categories* listed below.

1. Plans, Intentions, and Strategies
2. Organizations and Leadership

3. Operations
4. Infrastructure
5. Systems Acquisition and Implementation

Category 1: Plans, Intentions, and Strategies

Although the customer's initial question will likely be framed in terms of one of the categories, consider if it needs to be broadened to include other categories. The questions listed below will assist analysts in determining the answers needed to address this category.

1. *What is the target's **motivation**?*
2. *What is the target's **doctrine** (stated beliefs)?*
3. *What shapes the target's **perceptions**?*
4. *What are the target's **priorities**?*
5. *What are the target's **strategies**?*
6. *What is driving the target's **intent**?*

Category 2: Organization and Leadership

When a customer poses an intelligence question, there are often circumstances and conditions surrounding it the customer has not considered. Understanding the components listed below will assist analysts in determining the information needed to address this category.

1. The organizational structure
2. The identities of key actors
3. The roles of key actors
4. The purpose of key actors
5. Internal influences within the organization
6. External influences outside the organization

Category 3: Operations

As intelligence analysts gain more experience, they become increasingly aware that customers at times do not know exactly what they are asking for. This is especially true in the area of operations. Understanding the components listed below will assist analysts in determining the information needed to address this category.

1. Individual and organizational activities

2. Specific methods and implementations
3. The tools and tactics used
4. Spatial components (space / location)
5. Temporal components (time)

Category 4: Infrastructure

There are many times when a customer knows that an issue has potentially important implications but is unsure of the form the issue may take. This is true when considering the complex category of infrastructure. Understanding the components listed below will assist analysts in determining the information needed to address this category.

1. Utilities
2. Bandwidth
3. Computational power
4. Datacenters and/or servers
5. Transportation networks
6. Command and control centers
7. Safe houses (areas of safe operation)

Category 5: Systems Acquisition and Implementation

Having insight into a customer's priorities makes it easier to establish valid assumptions regarding their requirements for intelligence issues. Insight into the target's system acquisition and implementation capabilities makes it easier for analysts to address these priorities. Understanding the components listed below will assist analysts in determining the information needed to address this category.

1. Research
2. Development
3. Training
4. Capabilities
5. Readiness
6. Technical parameters
7. Vulnerabilities and weaknesses

PMESII: Operational Values

Customers tend to think about actions. Intelligence analysts must think about both the adversary's actions and the likely impact of the customer's reactions. *PMESII operational values* assist the analyst by providing context and clarity to their intelligence efforts.

1. **Political** (responsibility and power)
2. **Military** (offensive capabilities)
3. **Economic** (individual and group actors)
4. **Social** (values, behaviors, and beliefs)
5. **Infrastructure** (services needed to function)
6. **Information** (the ability to collect, process, and act)

Overview: ASCOPE / PMESII Matrix

The *ASCOPE PMESII Matrix* allows analysts to identify known data, intelligence gaps, and focus on specific areas without losing the broader scenario view.

OPERATIONAL VALUES

ENVIRONMENTAL FACTORS

	P Political	M Military	E Economic	S Social	I Information	I Infrastructure
A Areas	Areas Political	Areas Military	Areas Economic	Areas Social	Areas Information	Areas Infrastructure
S Structures	Structures Political	Structures Military	Structures Economic	Structures Social	Structures Information	Structures Infrastructure
C Capabilities	Capabilities Political	Capabilities Military	Capabilities Economic	Capabilities Social	Capabilities Information	Capabilities Infrastructure
O Operations	Operations Political	Operations Military	Operations Economic	Operations Social	Operations Information	Operations Infrastructure
P People	People Political	People Military	People Economic	People Social	People Information	People Infrastructure
E Events	Events Political	Events Military	Events Economic	Events Social	Events Information	Events Infrastructure

ASCOPE: Environmental Factors

The *ASCOPE* categories are environmental factors used to analyze the cultural and social conditions in which the operational variables (PMESII) exist.

1. **Areas**

2. **Structures**
3. **Capabilities**
4. **Organizations**
5. **People**
6. **Events**

Example: ASCOPE / PMESII Matrix for APT33

This method considers the perspective of people, culture, and society in an intelligence effort to provide a holistic view of the tactical operating environment.

OPERATIONAL VALUES

ENVIRONMENTAL FACTORS

	P Political	M Military	E Economic	S Social	I Information	I Infrastructure
A Areas	Regional Boundaries	Offensive, Defensive, and Current Operations		Dark Web, BBS, and Chat	Social Media and the Dark Web	Internet Service Providers
S Structures			Crypto Currency Exchanges			
C Capabilities	Local and National Political Influence	Strengths, Weaknesses, Opportunities, and Threats	Access to Funding Sources	Direct and Local Global (WEB)	Education, Training, and Technology	Ability to Maintain and Service Technology
O Organizations	Criminal, Legal, Political	Governmental and Nonstate	State Sponsored and Criminal	The Black Hat Consortium	TV Media and News Groups	Local Tech Companies
P People	Government and Provincial Leaders	Local, Proxy, National, and International				Programmers, Developers, Hardware, and HUMINT
E Events	Rallies, Protests, and Meetings	Exploit Events and Operations			Propaganda Recruiting Fundraising Campaigns	

The Key Assumptions Check

A *key assumption* is a hypothesis that has been accepted as true and will be a part of the problem definition or the final intelligence assessment product. A risk posed to the analysis process is that assumptions are not questioned in the beginning of the process. If these assumptions are not doubted at the start of the effort, they may become accepted as fact (incorrectly) going forward.

Analysts must identify any key assumptions presented and question their validity and relevance. All accepted assumptions must be explicitly stated, and a relevancy check must be conducted once they have been identified.

The Key Assumptions Check: Questions to Ask

Customers can sometimes choose not to be informed by intelligence analysis on a given issue or to ignore challenges to their assumptions and/or opinions. It is the analyst's responsibility to recognize this reality and guide the customer accordingly by asking the right questions. Listed below are several questions analysts will find useful for confirming the validity and relevance of any key assumptions put forward by the customer.

1. *How much confidence exists in the assumption?*
2. *How is the confidence level justified?*
3. *What forces or factors can undermine the assumption?*
4. *Could an assumption that was true in the past now be false?*
5. *Are there any penalties for a wrong assumption and, if so, what are they?*
6. *Is the key assumption in question a key factor in intelligence planning?*
7. *Has the process of vetting existing key assumptions revealed any new assumptions?*

Issue Decomposition

Issue decomposition is the technique for taking a seemingly intractable problem and breaking is down into a series of manageable sub-problems. It is presented as a conceptual framework and used to define issues and problems in detail. Depending on the intelligence agencies, this process is also referred to as *Strategies-to-Task*, *Problem Decomposition*, and *Decomposition and Visualization*.

Example Issue Decomposition: Cyber Threat

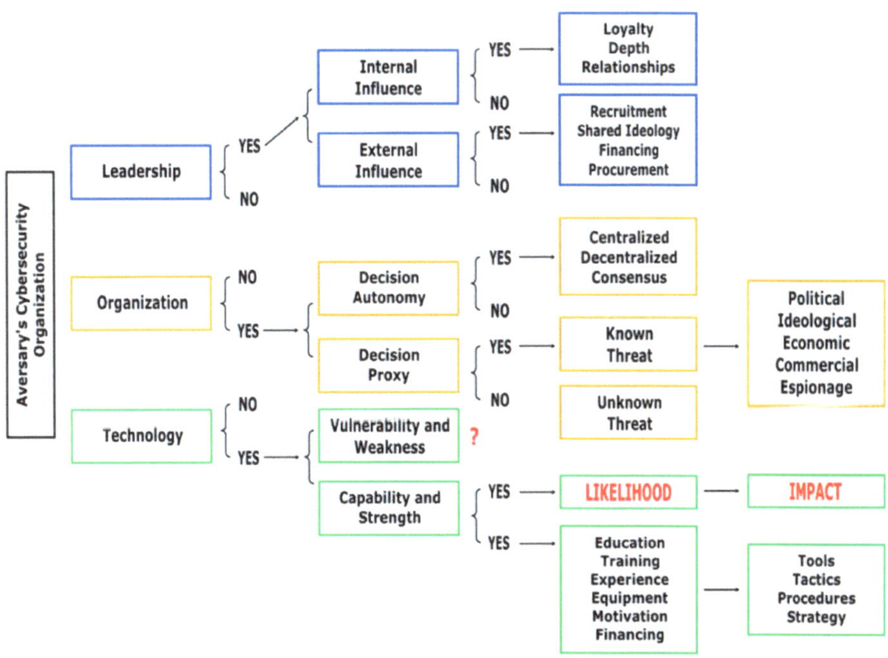

Issue Decomposition Taxonomy

A *taxonomy* is a classification system in which objects are arranged into natural or related groups based on a factor common to each object in the group. The process starts from the *issue definition statement* which provides specific details about the problem. The process also serves to define intelligence needs.

The intelligence issues are arranged from top level down to the specific task level. The subtasks are referred to as *key intelligence questions*. This approach has an intuitive appeal for intelligence analysts as humans tend to think about issues hierarchically by nature. This taxonomy also allows for an effective presentation of a final intelligence product to the customer.

CHAPTER 06

Determining the Root Cause of Intelligence Issues

Knowledge Assessment Questions

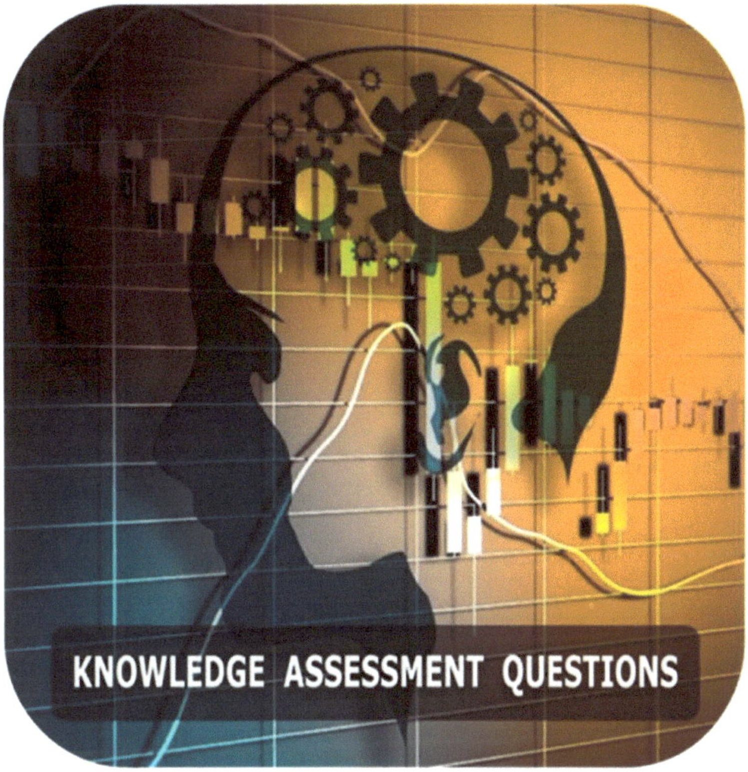

NOTICE TO STUDENTS:

The following knowledge assessment questions are presented in true / false, multiple choice, and fill-in-the-blank formats. The correct answers are provided in an Answer Key at the end of Chapter 14. These questions may or may not be presented on the quizzes and/or certification exam given by the instructor of this course.

Knowledge Assessment Questions

1) On large and complex intelligence projects, the __________ statement will itself become a formal product to be disseminated to stakeholders.

A. Qualitative

B. Parameter

C. Quantitative

D. Implementation

2) When analysts receive the initial tasking (from a customer of self-initiated) time pressures may tempt them to give the __________ definition minimal consideration.

A. Reporting

B. Collection

C. Issue

D. Reasoning

3) What choice listed below is a factor to be considered by an intelligence analyst when defining the requirements for the *Systems Acquisition and Implementation* issue category?

A. Key Actors

B. Implementations

C. Utilities

D. Readiness

4) A __________ is a classification system in which objects are arranged into natural or related groups based on a factor common to each object in the group.

A. Taxonomy

B. Doctrine

C. Ontology

D. Ideology

5) A __________ is a hypothesis that has been accepted as true and will be a part of the problem definition or the final intelligence assessment product.

A. Logical Fallacy

B. Key Assumption

C. Issue Definition

D. Cultural Bias

True / False

6) All intelligence analysis efforts begin with an issue or problem definition which aligns with the customer's interest about the target.

A. True

B. False

7) Target deconstruction is the technique for taking a seemingly intractable problem and breaking is down into a series of manageable sub-problems.

A. True

B. False

8) Having insight into an analyst's priorities makes it easier to establish valid assumptions regarding their requirements for intelligence issues.

A. True

B. False

9) A well-defined statement of the problem (the result of these answers) still requires two additional steps: issue definition and issue decomposition.

A. True

B. False

10) The PMESII categories are environmental factors used to analyze the cultural and social conditions in which the operational variables (ASCOPE) exist.

A. True

B. False

CHAPTER 07

Creating Targeting Models and Intelligence Profiles

Lecture Summary

KEY KNOWLEDGE POINTS

Modeling the Intelligence Target

Intelligence Modeling Concepts

Creating Intelligence Profiles

The Model Hierarchy

Creating a Target Framework

Target Frameworks: Deconstructing Elements

Modeling the Intelligence Target

Developing an intelligence assessment leads to the creation of a *model* (or set of models) of the target if the models do not already exist. A model is a replica or representation of an idea, object, or system. All models begin in the human mind and are essential when dealing with complex targets. Models also serve as an organizing constraint for analysts, restricting them to focusing on relevant facts, hypothesis, and assumptions.

Intelligence Modeling Concepts

Instead of interacting with the real system, an analyst can create a model which corresponds in specific ways to the actual target. Models used in intelligence tend to fall within one or more of the five categories listed below.

1. **Physical Models**: A tangible representation of a thing
2. **Conceptual Models**: Creations of the mind for abstraction
3. **Descriptive Models**: Describing the representation
4. **Normative Models**: Prescribing a course of action
5. **Decision-Support Models**: A comparison among *Alternative Competing Hypotheses* (ACHs)

The Model Hierarchy

Visualization involves transforming an intelligence product into graphic, pictorial, or multimedia forms so the brain can process and understand complex data sets.

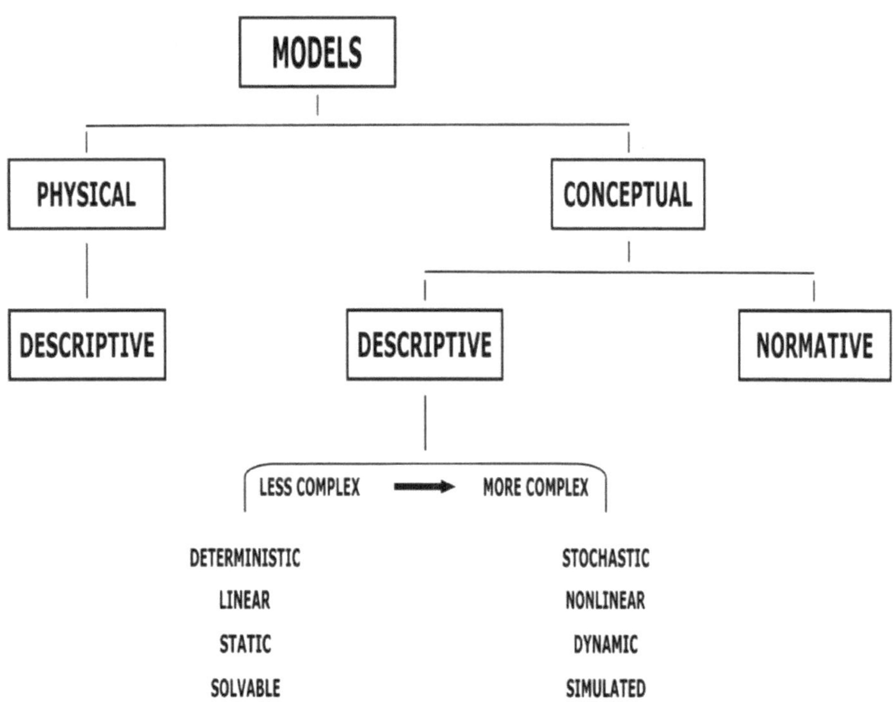

Descriptive Models

Modeling does not always require a computer, and many useful models that analysts can utilize to solve complex issues exist only on paper. Without a device to capture the full range of thinking and creativity that occurs in target modeling, analysts would have to keep track of too many details. The eight examples listed below are visual representations of models listed in the model hierarchy.

1. **Deterministic Models**: Relations are known and specified
2. **Stochastic Models**: Contain uncertainties and probabilities
3. **Linear Models**: Linear equations are used to describe relationships
4. **Nonlinear Models**: Use several types of mathematical functions
5. **Static Models**: Ignore time-based variances (snapshots in time)
6. **Dynamic Models**: Include spatial and temporal aspects and characteristics
7. **Solvable Models**: Analysis can be used to find probable answers
8. **Simulated Models**: "If Then" scenarios with multiple variables to predict outcomes

Comparative Models

Comparative techniques are a simple but useful form of modeling available to analysts that do not require sophisticated simulation exercises. They are based on the concept of *benchmarking*, comparing "us" to "them. The tend to be presented as either *parallel list* or *matrices*.

Example: APT Comparative Threat Matrix

Threat Matrix: APT Comparative Analysis

Risk Values 1-20	**Known TTPs** Documented	**APT 28** Fancy Bear	**APT 32** OceanLotus	**APT 33** Elfin Team	**APT 41** Wicked Panda
1-20	Targets Petrochemical and Energy Industries	0	0	20	1
1-20	Targets Regions of Organizational Operation	0	0	20	1
1-20	Capable of Executing Zero-Day Exploits	2	4	5	6
1-20	Uses Boot or Auto-Start Execution	15	15	10	10
1-20	Targets Non-Standard Ports and/or Protocols	15	15	10	1
1-20	Uses Encryption for C2 and Data Exfiltration	15	15	15	10
1-20	Targets Windows Servers and File Sharing Systems	15	15	15	10
1-20	Uses Spear Phishing to Compromise Accounts	20	20	20	7
Total Value 1-160		**82**	**84**	**115**	**46**

Small Number Documented	Moderate Number Documented	Significant Number Documented
Risk Values		
1-4 / **Low Probability**	5-10 / **Moderate Probability**	11-20 / **High Probability**

Parallel lists are used for comparing alternative solutions and are a key component for analysts engaged in comparative intelligence analysis. Matrices – which can be qualitative or quantitative – are an effective was of presenting visual results to customers.

Comparative Modeling: Disadvantages

One disadvantage of comparative modeling is that analysts may be inclined to rely on familiar models (cultural bias) instead of those of the target's culture. The most prevalent culture bias is referred to as *mirror imaging* which leads to inaccurate estimates and analysis. Comparative models must be specific to the culture. If not, a risk exists that strengths and weaknesses may be overlooked, as well as critical dependencies within the target network. *Parallel models* can help to overcome this risk in analysis as they support analytical benchmarking.

Creating Intelligence Profiles

The purpose of creating *intelligence profiles* is to help predict what the profiled person or group will most likely do in various circumstances and conditions. Intelligence profiles are models of individuals or groups, documenting a pattern of mental or behavioral traits. The traits are collectively referred to as the *modal personality*. These backgrounds – based on observed behavior and biographical data – create specific portraits of the target. These profiles can also predict the behavior of cyber criminals, as modal personalities exist equally online.

Pattern Models

Pattern modeling is a critical element of all intelligence as leaders, terrorists, and criminals have a modus operandi (a standard operational pattern). They are used to recognize patterns of activities such as departure or changes from the known state of the target. These models rely heavily on statistics and software to identify discrete trends that would be difficult to notice by human analysts. Also referred to as *Pattern-of-Life* (POL) modeling, they are useful in identifying and predicting future activity. These models can also be incorporated into nonvisual surveillance techniques such as reviewing browsing histories, phone calls, and financial transactions.

Creating a Target Framework

The generic target framework is the starting point for any analysis, as defining and deconstructing the target framework is the most important step in analysis. Once the system and issues of interest have been defined, a generic target framework is selected that falls into one of the three categories listed below.

1. **Structural System Models**: These include actors, objects, and relationships
2. **Functional System Models**: These models concentrate on achieved results
3. **Process System Models**: These models focus on interactions and their dynamics

Target Frameworks: Deconstructing Elements

Starting from the top-level view of the target, the analyst will *deconstruct* it into sub-categories (*elements*) and then sub-elements (all represented by models). This creates a visual picture of the target and commonly takes the form of a set of interrelated models. Target deconstruction serves many purposes, several of which are listed below.

1. Supports analyzing known intelligence
2. Identifies and closes gaps in knowledge
3. Helps to develop a finished intelligence product

4. Identifies what is known, unknown, and assumed.

Network Target Framework: Top Level

In the case of a network, target deconstruction includes understanding how the network functions and how the sub-elements interact with each other.

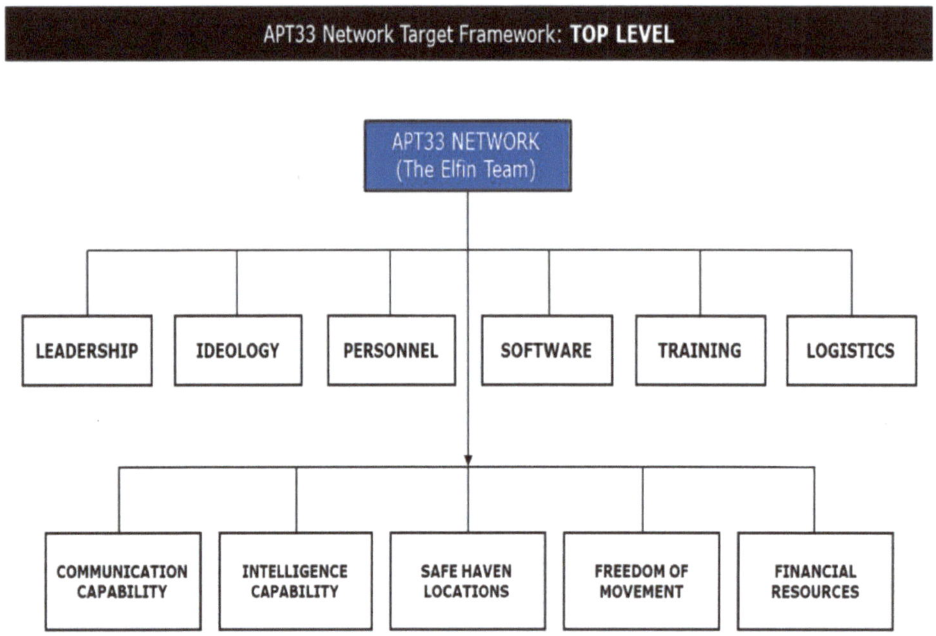

Network Target Framework: Deconstruction 01

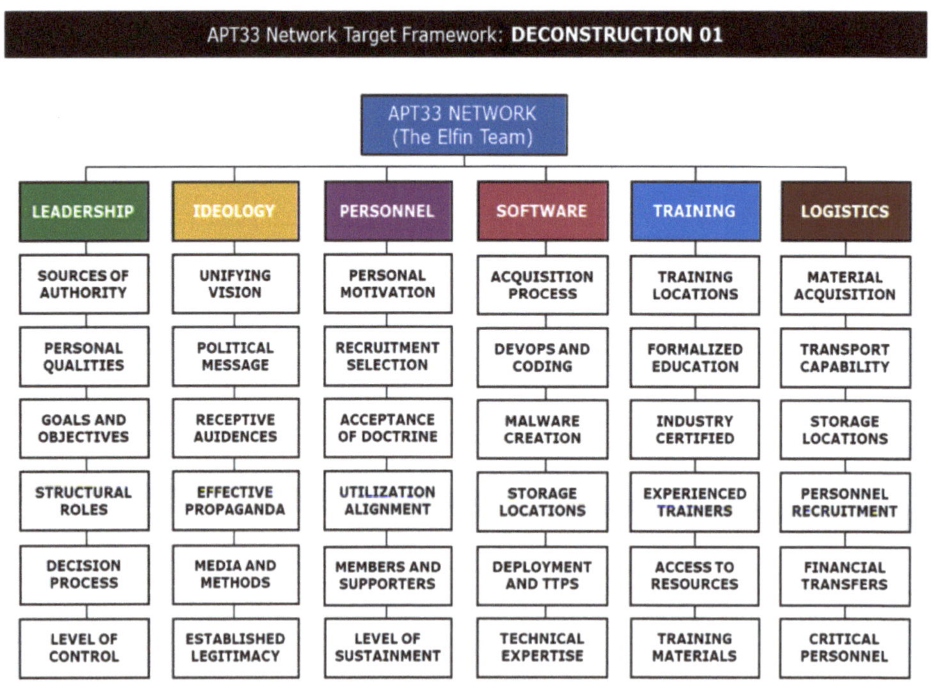

Network Target Framework: Deconstruction 02

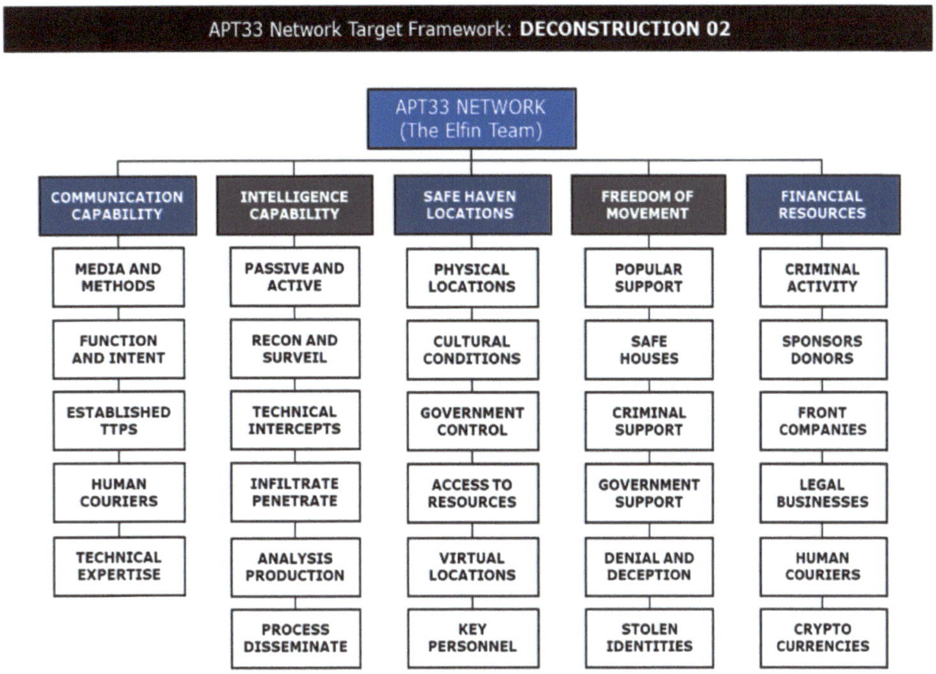

Alternative Frameworks

Alternative frameworks and their supporting models are valuable assets for analysts to use and an essential part of the intelligence analysis process. Analysts working to deconstruct issues should keep multiple frameworks in mind. To maximize the usefulness in identifying conflicting intelligence information multiple alternative models are needed.

The risk posed to analysts is that analytic pitfalls can undermine intelligence assessments and lead to a failure to consider these alternative scenarios. The three analytic pitfalls listed below are the most common an analyst will experience.

1. *Ego*: A summary rejection of dissenting views.
2. *Time*: Tight deadlines force analysts to make easy choices.
3. *Customers*: A tendency to choose the most favorable alternative instead of the most realistic option.

Competitive Frameworks

It is an established standard in intelligence that, if the organization can afford the resources, independent groups should provide competing analysis. Analysts are always dealing with uncertainty, and different analysts may arrive at different conclusions. Even customers may provide competing analysis. The process should be inclusive when defining frameworks – nothing should be summarily dismissed – as alternatives not offered or rejected without proper consideration may be the correct conclusion.

[THIS PAGE IS INTENTIONALLY LEFT BLANK.]

CHAPTER 07

Creating Targeting Models and Intelligence Profiles

Knowledge Assessment Questions

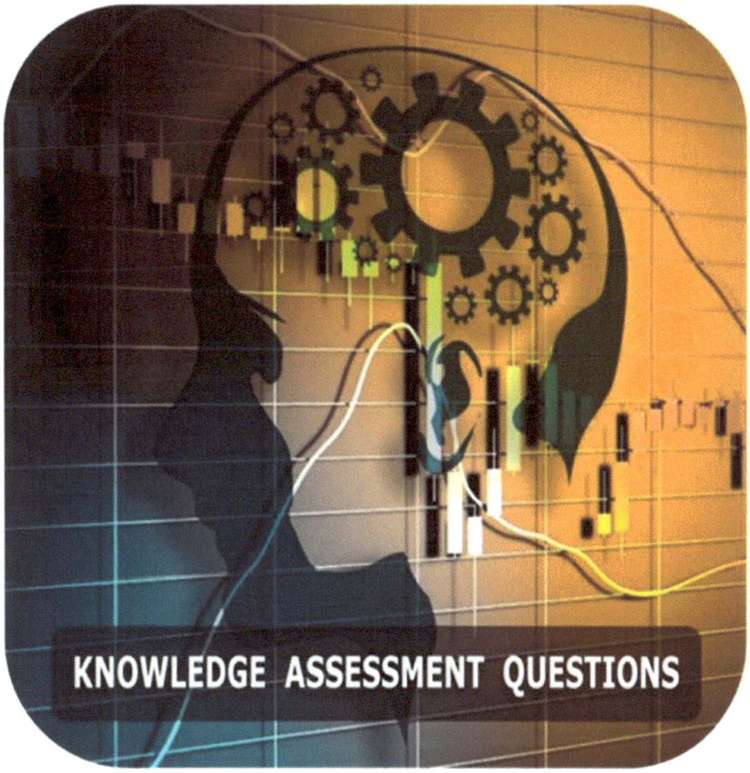

NOTICE TO STUDENTS:

The following knowledge assessment questions are presented in true / false, multiple choice, and fill-in-the-blank formats. The correct answers are provided in an Answer Key at the end of Chapter 14. These questions may or may not be presented on the quizzes and/or certification exam given by the instructor of this course.

Knowledge Assessment Questions

1) When working with intelligence modeling concepts, analysts typically use __________ models to prescribe a course of action for an issue.

A. Normative

B. Conceptual

C. Descriptive

D. Physical

2) When working with descriptive models, analysts use __________ models when the issue or problem contains uncertainties or probabilities.

A. Deterministic

B. Stochastic

C. Linear

D. Nonlinear

3) It is an established standard in intelligence that, if the organization can afford the resources, independent groups should provide __________ analysis.

A. Supporting

B. Accurate

C. Competing

D. Speculative

4) Analysts creating target frameworks for analytical work use __________ system models when describing actors, objects, or relationships.

A. Elemental

B. Process

C. Functional

D. Structural

5) The purpose of creating intelligence __________ is to help predict what the profiled person or group will most likely do in various circumstances and conditions.

A. Profiles

B. Cases

C. Models

D. Frameworks

True / False

6) Modeling does not always require a computer, and many useful models that analysts can utilize to solve complex issues exist only on paper.

A. True

B. False

7) Coordinating techniques are a simple but useful form of modeling available to analysts that do not require sophisticated simulation exercises.

A. True

B. False

8) Competitive modeling is a critical element of all intelligence as leaders, terrorists, and criminals have a modus operandi (a standard operational pattern).

A. True

B. False

9) The generic target framework is the starting point for any analysis, as defining and deconstructing the target framework is the most important step in analysis.

A. True

B. False

10) Starting from the low-level view of the target, the analyst will deconstruct it into mid-categories (elements) and then higher-elements (all represented by models).

A. True

B. False

CHAPTER 08

Evaluating Intelligence Sources and Evidence

Lecture Summary

KEY KNOWLEDGE POINTS

Issue Decomposition and the Target Framework

Analyzing Existing Intelligence

Sources of Raw Intelligence

Evaluating the Credentials of Evidence

Combining Types of Evidence

The Role of Information Technology

Issue Decomposition and the Target Framework

The *issue decomposition* and the *target framework* are separate concepts and separate models, but they are closely related to each other. Decomposition is asking the right questions. Target models provide the analyst with the correct context. Although similar in structure, the two models are separate for the reasons listed below.

1. The models come from different sources.
2. The models change in different ways.
3. An issue may address a portion of the model.
4. A model may serve several issues.

Analyzing Existing Intelligence

In intelligence analysis, *collation* is the organizing of relevant information in a coherent way, taking both the source and the context into consideration. The analyst must evaluate the intelligence for relevance and credibility, then collate the intelligence into the target model using the steps listed below.

1. Review the existing finished intelligence.
2. Examine the existing raw intelligence.
3. Acquire new raw intelligence if it is available.
4. Evaluate the new raw intelligence.
5. Combine the intelligence into the generic target framework.

Existing Finished Intelligence

Prior to starting an intelligence collection effort, analysts should ensure that they are already aware of information already collected on a target. The analyst starts with existing reports on file as most intelligence issues have already been addressed to some degree.

The first step begins with a literature search to establish the current state of knowledge and identify controversies and disagreements which may exist. Do not automatically accept existing intelligence as fact; conduct a key assumption check on the model to validate its relevance. Finally, review the existing raw intelligence data with the same level of scrutiny.

Sources of Raw Intelligence

Successful analysis requires an understanding of the sources of intelligence, and the nature and limitations available from those sources. Intelligence has two primary source types: *literal* and *nonliteral*. *Literal intelligence* exists in a form used for communication – understood by analysts and customers – using material requiring no special exploitation. Nonliteral intelligence requires special processing to be understood, such as imagery and foreign language communication. It is not a perfectly clean distinction and is treated differently depending on the source.

Chapter 08: *Evaluating Intelligence Sources and Evidence*

Evaluating Evidence

The fundamental objective in weighing intelligence evidence is determining its credibility; the analyst must evaluate it for completeness and veracity. This process involves subjective judgments. A *quality of information check* is used to evaluate the *source*, *channel*, and *evidence*. It is best to adhere to the principles of *Occam's Razor* to explain the evidence using the fewest hypotheses possible. However, do not be restricted by the principle. Answers to questions may be highly complex and not expanding beyond *Occam's Razor* presents the risk of oversimplification.

Evaluating the Source

It is possible for analysts to make information fit almost any desired conclusion, especially by selectively disregarding conflicting or inconvenient facts. The analyst should accept nothing at face value and evaluate the origin of the evidence carefully, being aware of potential motives of the source. The three questions listed below can provide a quick (and superficial) evaluation of the source.

1. *Is the source competent?*
2. *Does the source have required access to obtain the information?*
3. *Does the source have a vested interest in the outcome?*

Evaluating the Communication Channel

Over a long chain of human communication (especially in bureaucracies), the received message may bear little resemblance to the information originally sent. The message tends to get distorted in transit as randomness increases in any system. Accuracy decreases as the length of the channel increases and as the number of nodes in the channel increases. Accuracy will also depend on whether the source was *primary* (firsthand information) or *secondary* (using intermediaries). Entropy will always have an impact on intelligence and contribute to "*The Creeping Validity Problem*."

Instead of a "primary-versus-secondary" distinction, it is more important for the intelligence analyst to consider the communication channel itself. Analysts should ask the questions listed below to provide clarification and verification of the channel used during the process.

1. *What was the channel?*
2. *Was the information provided intentionally?*
3. *What part of the information is true?*
4. *Could the information be part of a deception and denial effort?*
5. *Could the information be a signal to the adversary?*
6. *If the information is a signal, what is the reason?*

Source Analysis

The synergistic effect of combining information from multiple sources serves to strengthen conclusions and increase the team's confidence in them. Analysts conducting an analysis of the source should consider the past reporting history and the psychological profile. An incomplete picture can be misleading – counterarguments from sources tend to be avoided – and the analyst must take the necessary steps to find missing evidence to complete the picture.

Source Reliability

The *source reliability* is rated between "A" (complete reliability) to "E" (invalid information), with "F" for a source without a sufficient history to rate accurately. Listed below is an example of a common system used to rate source reliability.

1. **A**: *Reliable* (a history of complete reliability)
2. **B**: *Usually Reliable* (a stable source, but minor doubts exist)
3. **C**: *Fairly Reliable* (the amount of doubt and confidence are equal)
4. **D**: *Not Usually Reliable* (there are significant doubts regarding source information)
5. **E**: *Unreliable* (the source lacks authenticity or is completely invalid)
6. **F**: *Unknown* (there is insufficient information to decide)

Information Reliability

The information reliability is rated between "1" (confirmed) to "5" (improbable), with a "6" for information that cannot be confidently evaluated by the analyst. Listed below is an example of a common system used to rate information reliability.

1. **1**: *Confirmed* (by independent sources)
2. **2**: *Probably True* (logical, but not confirmed)
3. **3**: *Possibly True* (reasonably logical, but not confirmed)
4. **4**: *Doubtful* (not logical and not confirmed)
5. **5**: *Improbable* (not logical with contradicting information)
6. **6**: *Unknown* (the validity of the information cannot be determined)

Vested Interest and Bias

Analysts occasionally encounter the "professional source" who sells information to the highest bidder and makes it as interesting as possible. Government agencies have similar problems. Officials may distort or conceal facts to support policy decisions and protect personal interests.

They may also exaggerate the importance of the situation to encourage additional government funding. Analysts should be wary of existing reports they are provided with as they may be slanted to prove or disprove a preferred position.

Chapter 08: *Evaluating Intelligence Sources and Evidence*

Evaluating the Credentials of Evidence

Credibility refers to the extent to which an analyst can believe something, and relevance means the information has value in addressing an issue or question. A *fact* is verified information known to exist. *Direct information* is likely factual because of the credibility of the source. *Indirect information* is provided by intermediaries and the veracity of the truth has not been established. Analysts should be aware that direct information may be misleading if it is used in denial and deception operations by the adversary.

Evaluating Evidence: Disadvantages

Customers place priority on evidence they collect themselves; a mistake that leadership can make easily which causes them to be vulnerable to deception. Listed below are common mistakes that may undermine the ability to evaluate evidence effectively.

1. Vividness weighting
2. Weighing based on the source
3. Favoring the most recent evidence
4. Favoring or disfavoring the unknown
5. Trusting hearsay
6. Relying on expert opinions
7. Premature closure
8. Philosophical disposition

Combining Types of Evidence

The intelligence analysis process involves combining disparate types of evidence and helps analysts to make qualitative judgments about conclusions to support. Listed below are four categories of evidence likely to be encountered by intelligence analysts.

1. **Divergent Evidence**: This category consists of two data sets that favor different conclusions.
2. **Convergent Evidence**: This category consists of two data sets that favor the same conclusion.
3. **Corroborative Redundancy**: This category consists of known information that is corroborated by multiple sources.
4. **Cumulative Redundancy**: This category consists of known information that is corroborated by multiple sources but is also enhanced with additional information provided by each source.

Analysis of Competing Hypotheses (ACHs)

It is often important to combine evidence and demonstrate the logical process of reaching a conclusion based on that evidence through *structured argumentation*. This can assist analytical judgment on important issues. Carefully weighing alternative conclusions helps to minimize cognitive limitations, assists in identifying a complete set of hypotheses, and supports the evaluation of consistent and inconsistent data. Viewing the target from an alternative perspective also allows analysts to analyze potential gaps and assists in the discovery and questioning of assumptions.

Combining Evidence with Bayesian Analysis

Bayesian analysis uses new data to modify previously estimated probabilities and can therefore be used to narrow the error bounds on estimates. Using this method, new information is evaluated individually. The new information combines *historical* with *subjective* assessment to determine the likelihood of an occurrence. The analyst can then attribute the observed data to specific causes. The difficulties associated with Bayesian analysis lie in the fact that it is difficult to teach and apply in practice.

The Role of Information Technology

A significant consequence of the information explosion is that analysts must now deal with "big data" when collating and analyzing intelligence products. Analysts face significant challenges dealing with large and complex data sets. Current database management tools tend to be ineffective for performing analytical tasks. Additionally, the amount of collected data tends to be too large to be stored and analyzed.

Information technology and specialized software provides analysts with the tools they need to address these challenges. Listed below are several well-known software packages used by analysts to assist them in their intelligence operations.

Palantir Technologies

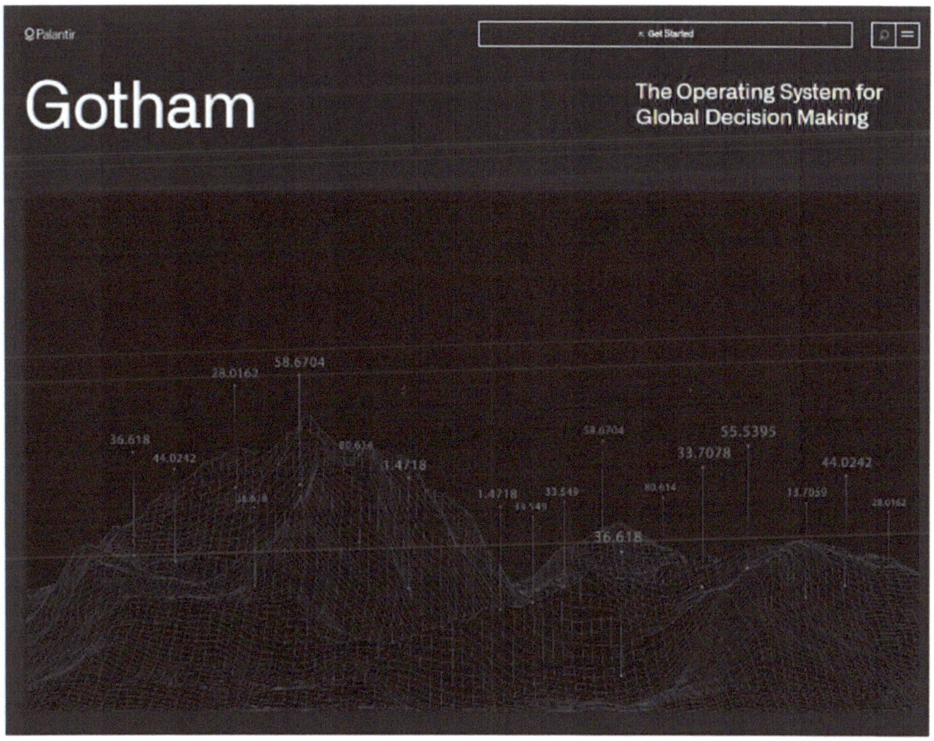

www.palantir.com

IBM i2 Analyst's Notebook

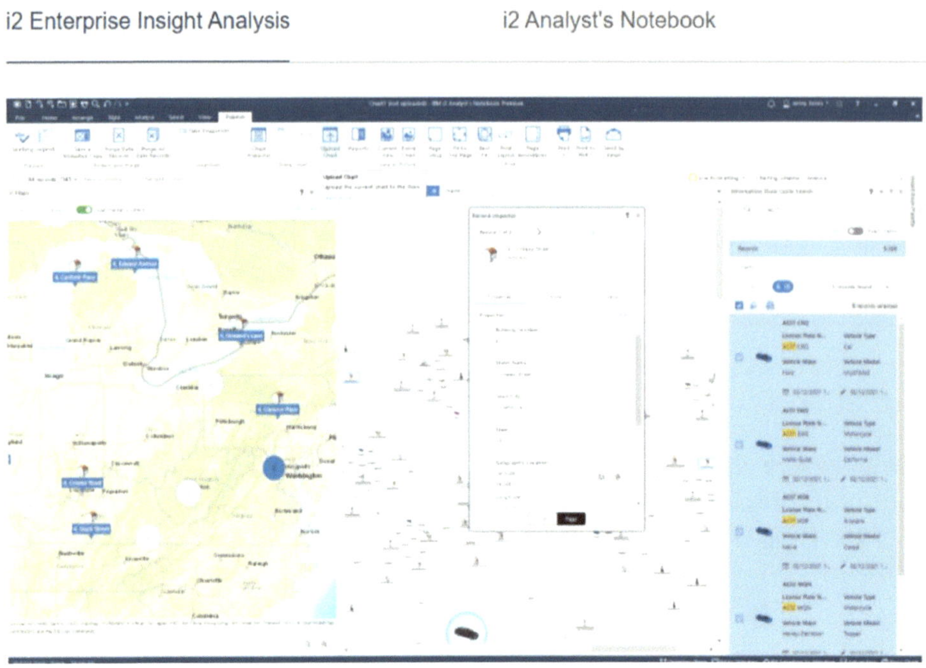

i2group.com

Visual Link

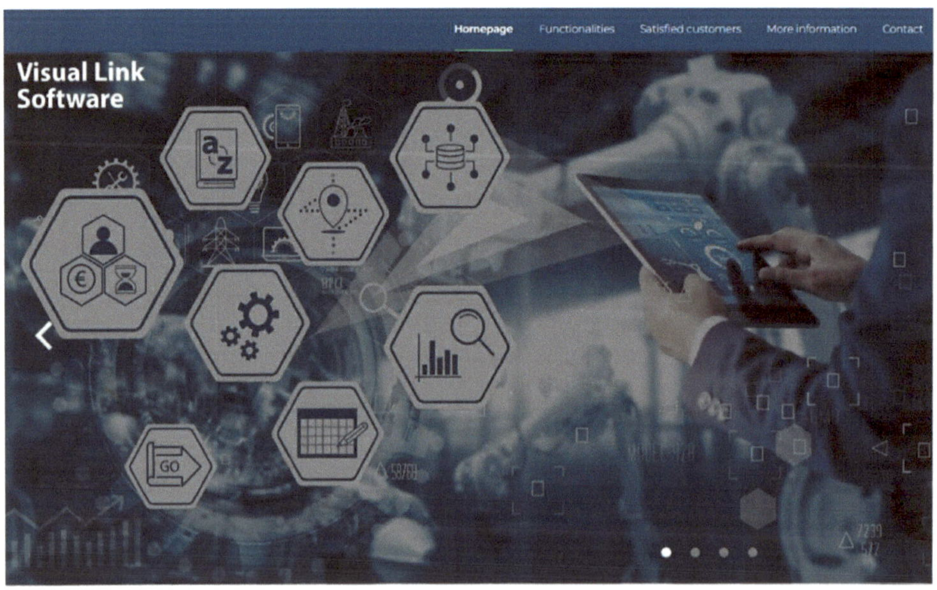

www.visuallink.nl

CHAPTER 08

Evaluating Intelligence Sources and Evidence

Knowledge Assessment Questions

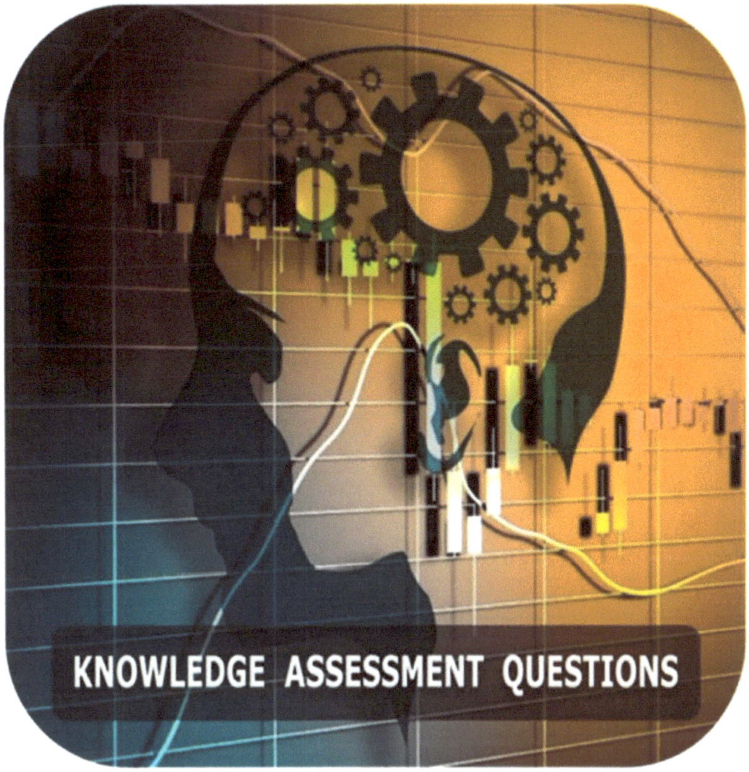

NOTICE TO STUDENTS:

The following knowledge assessment questions are presented in true / false, multiple choice, and fill-in-the-blank formats. The correct answers are provided in an Answer Key at the end of Chapter 14. These questions may or may not be presented on the quizzes and/or certification exam given by the instructor of this course.

Knowledge Assessment Questions

1) In intelligence analysis, __________ is the organizing of relevant information in a coherent way, taking both the source and the context into consideration.

A. Collection

B. Collation

C. Comparison

D. Corroboration

2) __________ analysis uses new data to modify previously estimated probabilities and can therefore be used to narrow the error bounds on estimates.

A. Theoretical

B. Historical

C. Bayesian

D. Substantive

3) Intelligence analysts working with __________ evidence are considering two data sets that favor different conclusions.

A. Corroborative

B. Convergent

C. Cumulative

D. Divergent

4) The fundamental objective in weighing intelligence evidence is determining its __________; the analyst must evaluate it for completeness and veracity.

A. Credibility

B. Source

C. Channel

D. Method

5) It is often important to combine evidence and demonstrate the logical process of reaching a conclusion based on that evidence through __________.

A. Literal Intelligence

B. Structured Argumentation

C. Professional Sources

D. Occam's Razor

True / False

6) A significant consequence of the information explosion is that analysts must now deal with "big data" when collating and analyzing intelligence products.

A. True

B. False

7) The intelligence analysis process involves combining identical types of evidence and helps analysts to make qualitative judgments about conclusions to support.

A. True

B. False

8) Prior to starting an intelligence collection effort, analysts should ensure that they are already aware of information already collected on a hypothesis.

A. True

B. False

9) The issue decomposition and the target framework are separate concepts and separate models, but they are closely related to each other.

A. True

B. False

10) The information reliability is rated between "A" (complete reliability) to "E" (invalid information), with "F" for a source without a sufficient history to rate accurately.

A. True

B. False

CHAPTER 09

Predicting the Adversary's Planning and Operations

Lecture Summary

KEY KNOWLEDGE POINTS

Anticipatory Analysis

Background Forces

Anticipatory Analysis Methodology

Extrapolation, Projection, and Forecasting

Influence Tree and Influence Net Diagrams

High-Impact / Low-Probability Analysis

Anticipatory Analysis

Estimating the future state of a target can only be done using probabilities, so the risk of a mistake is greater than when analysts rely on established facts. Anticipatory analysis requires a different conceptual framework referred to as *probabilistic analysis*, a method that requires highly structured thinking. It presents a view of what is likely to happen and takes the form of an *intelligence estimate*. The three primary goals of this analysis are listed below.

1. It characterizes and reduces uncertainty.
2. It reveals several potential outcomes.
3. It enhances overall resiliency and adaptability.

Most intelligence customers tend to focus on current situations, and those who take a strategic view are inclined to rely on their own predictive abilities. However, customers will accept anticipatory analysis if it is well-reasoned, understandable, and the logical development can be understood. Analysts must apply a formal methodology so the customer can see the basis for the conclusions. The focus of the analysis is to identify forces likely to affect the target, a process referred to as "*Outside In Thinking*."

The most important contribution at the start of an estimate is the assessment of factors that will shape future events and the future state of the target model. The four steps listed below will assist the analyst in accomplishing this task.

1. *List all external key factors*. The first set of factors should be those over which little influence can be exerted.
2. *List key fact that can be influenced*. This list includes the behavior of both adversaries and allies.
3. *Assess all the external key factors*. Identify the impact the factors may have on the analytic issue.
4. *Determine if the factors will have an impact*. The assessment of the analyst should be based on all available evidence.

Background Forces

Some forces are both broad and subtle; they operate as a background process on most targets and need to be considered by the analyst. The four background forces of most concern to the analyst are listed below.

1. **Inertia**: The tendency to resist change
2. **Opposition**: Opposing or resistive forces
3. **Contamination**: The degradation of any external factor or force
4. **Feedback**: The ability of the system to adapt, learn, and change

Synergistic Considerations

It is essential that the intelligence analyst develop an understanding of a broad range of concepts to function in a multidisciplinary environment. *Synergy* is a term used to describe a whole that is more than the sum of its parts. Synergy itself is not a force, but the result of the interaction of forces, cooperative efforts, and alliances. Synergy has a *force multiplier effect* in models.

Causal Models

Causal models depict elements of the target, elements affecting the target, and describe how those elements are related by probabilities. These models are composed of *nodes*. The nodes tend to be variables in which the relationship can be positive or negative. They use probability laws, distributions, and must meet the criteria listed below.

1. The time sequence must be proper.
2. There must be a *Theory of Causality*.
3. The relationship between the nodes cannot be questionable.
4. The analyst must prove covariation between the variables exists.

Anticipatory Analysis Methodology

In addition to taking a multidisciplinary approach, the second critical trait of a successful predictive analyst is the ability to establish proper historical context. This methodology uses past and current target models to help assess the potential future state of the target. The five steps listed below are the basic steps performed by the analyst when using this methodology.

1. Identify all forces acting on the entity.
2. Identify and associate all likely future forces.
3. Predict a range of like changes on the target.
4. Consider the impact of old and new forces over time.
5. Develop and state an indicator of confidence in the analytical judgments.

The Estimative Approach

The methodology (conceptual framework) of anticipatory intelligence relies on three predictive mechanisms: *extrapolation*, *projection*, and *forecasting*. Listed below are brief explanations of each mechanism.

1. **Extrapolation**: Assumes the forces acting on the target do not change.
2. **Projection**: Assumes the forces acting on the target do change.
3. **Forecasting**: Assumes the forces on a target do change and that new forces will be added.

The Estimative Methodology: Visual Representation

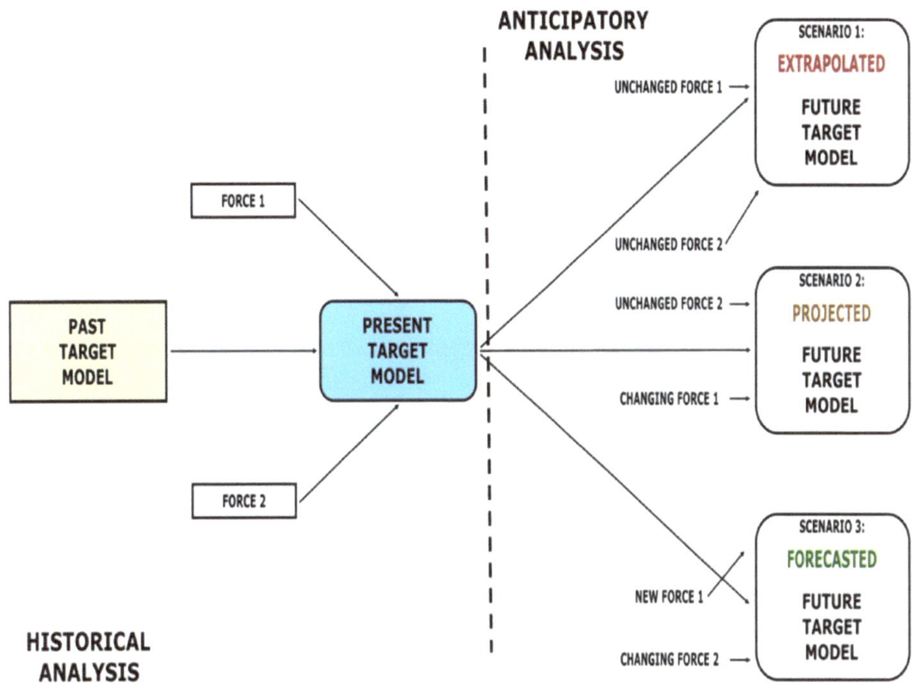

Extrapolation

An *extrapolation* is a statement, based only on past observations, of what is expected to happen. It is the most conservative method of prediction. This method is used when there is little uncertainty about the current state and the analyst has a high level of confidence in the forces acting on the target. New information is assigned low priority. It is typically accurate in the short-term assuming the analyst has an accurate starting point and knowledge of the direction of event movement.

Projection

Projection predicts a range of likely future outcomes based on the assumption that forces that have affected the target in the past will change. This method makes use of two analytic techniques: *force analysis* and *probabilistic reasoning*. *Force analysis* is a qualitative study of the forces acting on the target. *Probabilistic reasoning* is a systematic method of reasoning using subjective estimates of probabilities, making estimates explicit and consistent.

Generating Alternative Outcomes

The first step in probabilistic reasoning is formally stating that more than one outcome is possible (*alternatives*) by listing all foreseeable possible scenarios. These outcomes are defined as *alternative scenarios*. They must be mutually exclusive and exhaustive. Listing potential outcomes can expand analytic perspectives and generate useful insights into intelligence issues. Listing all relevant outcomes helps to prevent *premature closure* and helps to add structure to complicated issues.

Influence Tree Diagrams

Influence tree diagrams represent a systematic approach to force analysis and present a unique form of causal model for the intelligence analyst and customer. As an example, a customer requires an estimate of outcomes for APT33 (Elfin Team), and the analyst defines the three possible outcomes listed below.

1. APT33 successfully compromises a target.
2. A defense against an APT33 attack is successful.
3. The status of APT33 remains unchanged (no attack occurs).

The analyst must describe the forces affecting possible outcomes – referred to as driver events or linchpin premises – based on explicit and well-reasoned assumptions.

Example: Influence Tree Diagram

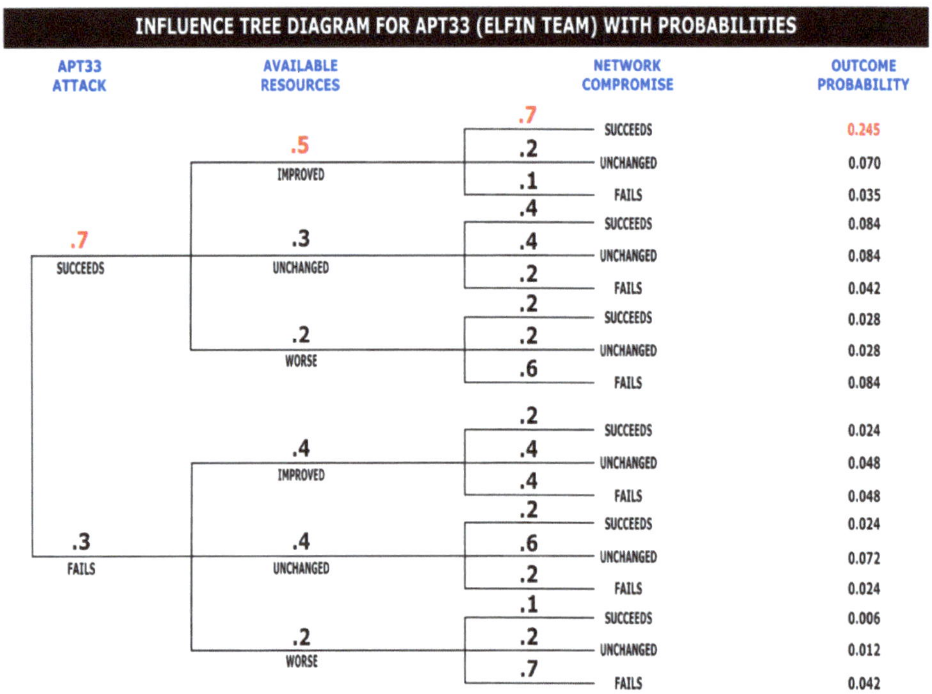

Influence Net Modeling Diagrams

Influence net modeling is a powerful tool for projection of complex target models where the influence tree would be too cumbersome for practical analytical use. This approach combines two methods of decision analysis: *Bayesian Inference Net Analysis* and *Influence Diagramming Techniques*. The analyst defines influence nodes using *arrowheads* (positive) and *circles* (negative).

This is an important tool in the target-centric approach to intelligence modeling. Additionally, the diagram can be shared with customers and encourages the customers to provide feedback and input.

Example: Influence Net Diagram

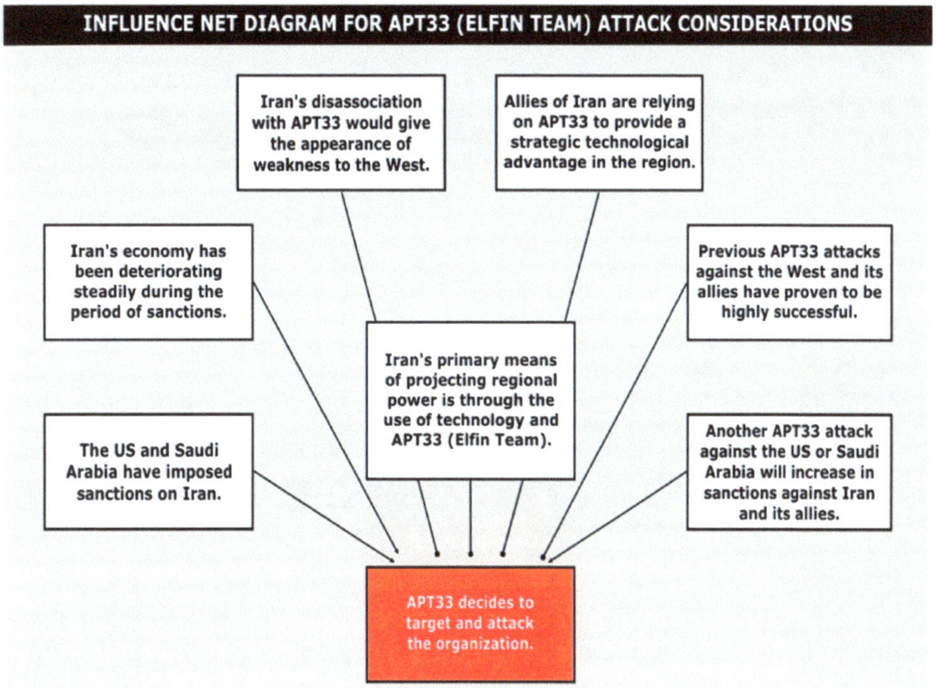

Forecasting

A primary objective of *forecasting* in intelligence operations is to define possible alternative outcomes of the target framework, not just the most likely ones. This technique is critical for effective strategic decisions. It attempts to identify new forces likely to act upon the target and considers all the possible impacts of these new developments.

It is critical that the analyst consider the background forces listed below when engaged in forecasting intelligence targets.

1. Inertia
2. Opposition
3. Synergy
4. Feedback

Because this technique is critical for strategic decisions, the use of high-level anticipatory analysis is preferred.

Evaluating Forecasts

Intelligence analysts should advise customers that forecasts are dynamic and need constant adjustment to be helpful in guiding thoughts and actions. Managing the customer's expectations is crucial. The seven factors listed below should be used by analysts to evaluate any forecast created for the customer.

1. *Clarity*
2. *Credibility*
3. *Plausibility*
4. *Relevance*
5. *Urgency*
6. *Comparative* Advantage
7. *Technology Quality*

High-Impact / Low-Probability Analysis

Projections and forecasts focus on the most likely outcomes, but customers also need to be aware of unlikely outcomes with severe adverse effects. This analysis – also referred to as a *demonstration scenario* – is used for sensitizing analysts and customers to the consequences of unlikely developments. These consequences typically arise from divergent phenomena and present themselves as unexpected and unpleasant surprises. The analysis describes how a specific development might begin, then continues to consider the plausibility and consequences of the development.

CHAPTER 09

Predicting the Adversary's Planning and Operations

Knowledge Assessment Questions

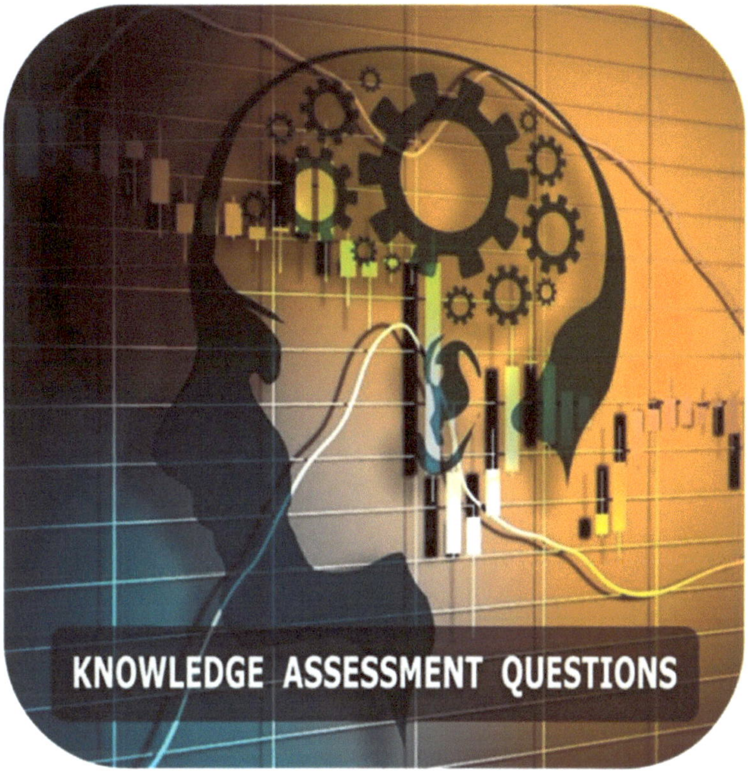

NOTICE TO STUDENTS:

The following knowledge assessment questions are presented in true / false, multiple choice, and fill-in-the-blank formats. The correct answers are provided in an Answer Key at the end of Chapter 14. These questions may or may not be presented on the quizzes and/or certification exam given by the instructor of this course.

Knowledge Assessment Questions

1) Analysts conducting anticipatory analysis must consider __________, a background force that addresses the tendency to resist change.

A. Opposition

B. Inertia

C. Contamination

D. Feedback

2) __________ is an estimative approach used by analysts in anticipatory analysis that assumes forces affecting the target do change.

A. Extrapolation

B. Forecasting

C. Projection

D. Prediction

3) __________ diagrams represent a systematic approach to force analysis and present a unique form of causal model for the intelligence analyst and customer.

A. Premature Closure

B. Influence Net

C. Demonstration Scenario

D. Influence Tree

4) Analysts conducting anticipatory analysis must consider __________, a background force that addresses the degradation of any external factor.

A. Contamination

B. Prediction

C. Opposition

D. Inertia

5) A primary objective of __________ in intelligence operations is to define possible alternative outcomes of the target framework, not just the most likely ones.

A. Projection

B. Forecasting

C. Extrapolation

D. Prediction

True / False

6) Estimating the future state of a target can only be done using probabilities, so the risk of a mistake is greater than when analysts rely on established facts.

A. True

B. False

7) Target models depict elements of the target, elements affecting the target, and describe how those elements are related by probabilities.

A. True

B. False

8) Intelligence analysts should advise customers that forecasts are static and need constant adjustment to be helpful in guiding thoughts and actions.

A. True

B. False

9) *Influence net* modeling is a powerful tool for projection of complex target models where the influence tree would be too cumbersome for practical analytical use.

A. True

B. False

10) The first step in probabilistic reasoning is formally stating that more than one outcome is possible (alternatives) by listing all foreseeable possible scenarios.

A. True

B. False

CHAPTER 10

Structured Techniques for Conducting Analysis

Lecture Summary

KEY KNOWLEDGE POINTS

Structured Analytic Techniques

Sorting Methodology

Matrices Methodology

Event Mapping Methodology

Event Tree Methodology

Subjective Probability Methodology

Structured Analytic Techniques (SATs)

The more important the issue, the more important that *structured analytic techniques* (SATs) become in the development of best judgment of the response. These techniques help the analyst make sense of complex problems by comparing and weighing pieces of information. They ensure the analyst focuses only on the issue at hand by forcing consideration of only one element at a time.

This method of consideration also reduces the opportunity to introduce biases and logical fallacies. SATs also help the analyst identify patterns and correlations, enhance data gathering and review, and provide an effective thinking model that allows alternative outcomes to be derived.

These techniques will improve analyses by making them more rigorous, provide ways to measure progress, and improve the presentation of the final product. Listed below are the six structured analytic techniques that will be addressed in this course.

1. *Matrices*
2. *Sorting*
3. *Threat Intentions Matrix*
4. *Event Mapping*
5. *Event Tress*
6. *Subjective Probability*

Matrices

Matrices are useful whenever there are more options or more intricate data than can be conceptualized at one time without a visual representation. This technique is useful for isolating critical data within abundant amounts of relevant information. Analysts can then focus on each option. This allows issue elements to be separated and categorized by type for comparison, helping to identify patterns and correlations. This technique is also an intermediate step in *link analysis*.

Matrices: Methodology

The two-dimensional design of matrices limits their use for collating data on complex issues and leaving out pertinent data will oversimplify issues. However, in those situation in which matrices apply, the steps listed below will assist analysts in using this technique effectively.

1. Draw a matrix using rows and columns.
2. Enter the desired range of data or criteria.
3. Note the relationships of data at grid intersections.
4. Review the initial hypothesis against the matrix.
5. Develop a new hypothesis if required.
6. Develop a matrix template for recurring topics.

7. Color-code the matrix results to aid in understanding and presentation.

Matrices: Example

The matrix below shows one method of cross-walking the operational variables memory aid *PMESII* and the environmental factors memory aid *ASCOPE*.

OPERATIONAL VALUES

ENVIRONMENTAL FACTORS

	P Political	M Military	E Economic	S Social	I Information	I Infrastructure
A Areas	Areas Political	Areas Military	Areas Economic	Areas Social	Areas Information	Areas Infrastructure
S Structures	Structures Political	Structures Military	Structures Economic	Structures Social	Structures Information	Structures Infrastructure
C Capabilities	Capabilities Political	Capabilities Military	Capabilities Economic	Capabilities Social	Capabilities Information	Capabilities Infrastructure
O Operations	Operations Political	Operations Military	Operations Economic	Operations Social	Operations Information	Operations Infrastructure
P People	People Political	People Military	People Economic	People Social	People Information	People Infrastructure
E Events	Events Political	Events Military	Events Economic	Events Social	Events Information	Events Infrastructure

Sorting

Sorting is effective when information elements can be broken out into categories for comparison using automated computer software. It is useful for reviewing massive datasets and multiple categories. This technique can present the analyst with possible trends and insights in information such as similarities, differences, and gaps. It is especially useful during initial data gathering and effective for initial hypothesis generation. Standardizing the data is imperative for sorting to function properly. The risk posed to analysts using this technique is that improper sorting can hide insights which should have been visible.

Sorting: Methodology

Sorting massive amounts of data can provide insights into trends or anomalies that warrant further investigation that may have otherwise gone unnoticed. The steps listed below will guide analysts through the general process of the sorting technique.

1. Review the chosen information categories.
2. Place the selected data into a spreadsheet.
3. List all the facts involved that will be used during the sorting process.

4. Use initial sorting to uncover possible patterns and groupings.

5. Identify grouping that illustrate the schema categories.

6. Group the data items according to schema.

7. Choose a single category and sort the data.

8. Review and re-review the sorted data to identify trends and/or anomalies.

Threat Intentions Matrix

The *threat intentions matrix* is an analytic technique used by intelligence analysts to efficiently look at information from the threat's point of view. It helps to mitigate analyst biases and provide insight into the threat's decision-making criteria. This technique develops clear indicators for each option which allows for specific collection planning.

To ensure this process is effective, all potential alternatives and options must be entered into the matrix – an effort that can be completed in a short amount of time – as the criteria used for this matrix does not tend to be extensive (description and depth). Once completed, better options and courses of action may become apparent to the analyst.

Threat Intentions Matrix: Methodology

It is necessary for the analyst using this technique to have knowledge of the motivation, goals, and objectives of the threat making the decisions. Armed with this knowledge, the steps listed below will guide the analyst in using this technique effectively.

1. Enter all decision **options** from the threat's point of view.

2. Enter the threat's **objectives** for each option.

3. Enter the perceived **benefits** of the threat's decision.

4. Enter the perceived **risks** associated with the threat's options.

5. Complete the implications column (a transition from the threat's view to the analyst's view).

6. Enter all perceived **implications** from the analyst's view.

7. Enter all perceived **indicators** from the analyst's view.

Example Threat Intentions Matrix: APT33

It is important to enter the information for every decision-making criteria in a column before moving to the next in order to mitigate analyst bias. An example of a threat intentions matrix focusing on APT33 – a recurring target throughout this course – is shown below.

Options	Objectives	Benefit	Risk	Implications	Indication
Organization's Public-Facing Website	Remove the ability to interact with the general public	Show of power and an embarrassment to the company	Mission failure, public embarrassment, and risk of sanctions	APT33 gains stature, Iran gains influence, new global sanctions	Ping scans, ping sweeps, and signs of potential DDoS
Organization's ICS / SCADA Control System	Destroy the ability to conduct critical operations	Show of power, media attention, and aid recruitment	Adequate training and maintaining security during the attack	APT33 gains stature, new sanctions, energy shortage	Increased phishing, actions consistent with ATT&CK Matrix
Organization's IAM / Password Database	Compromise the integrity of the credentialing system	Long-term access to sensitive data and accounts	Adequate training and maintaining security during the attack	APT33 gains stature, disrupt security, impact operations	Signs of network enumeration and increased phishing
Organization's Payroll / Finance Database	Disrupt the ability to conduct vital financial transactions	Show of power, disrupt operations and employee morale	Adequate training and maintaining security during the attack	APT33 gains stature, disrupt operations, loss of employees	Signs of network enumeration and spear phishing
Organization's Human Resource Database	Compromise and/or destroy employee personnel data	Acquire data on key employees for targeting efforts	Adequate training and maintaining security during the attack	APT33 gains stature, disrupt security, impact operations	Signs of network enumeration and spear phishing
Organization's Client Services Database	Compromise and/or destroy client business data	Show of power, disrupt operations and client confidence	Adequate training, security during attack and client response	APT33 gains stature, new sanctions, local economy impacted	Signs of network enumeration and attacks on clients
Organization's User Endpoint Devices	Establish a foothold for a large-scale network compromise	Acquire data about users and daily operations	Mission failure, public embarrassment, and impact recruitment	APT33 gains stature, disrupt security, impact operations	Increased phishing, event-based alerting, and help desk calls

Event Mapping

Event mapping uses a brainstorming diagram to represent the scenarios in the hypotheses linked around a central phrase representing the issue to be analyzed. This technique encourages a brainstorming approach and promotes creativity in generating new ideas. Additionally, it allows analysts to see associations not previously discovered. Elements are arranged intuitively by importance and arranged into *groups*, *branches*, and *areas*.

Participating in the process also tends to aid in the recall of existing memories within the analytical team. The risks posed to analysts using this technique is that it can become overly detailed if irrelevant events and ideas are included.

Event Mapping: Methodology

Use this technique when a nonlinear method is desired to generate, visualize, structure, and delineate the events in a scenario or related issue hypotheses. The steps listed below will guide the analyst in using this technique effectively.

1. Use a blank sheet of paper taped to a whiteboard.
2. List keywords or short phrases that represent comprehensive ideas.
3. List ideas as they occur and place them wherever they logically fit on the map.
4. Do not prejudge ideas, use restraint, or summarily dismiss ideas as irrelevant.
5. Develop potential directions for the topics.
6. Increase the detail as the map expands.
7. Use arrows to show links between events.
8. Provide context for the associations by notating the links.

Example Event Mapping: APT33 Attack

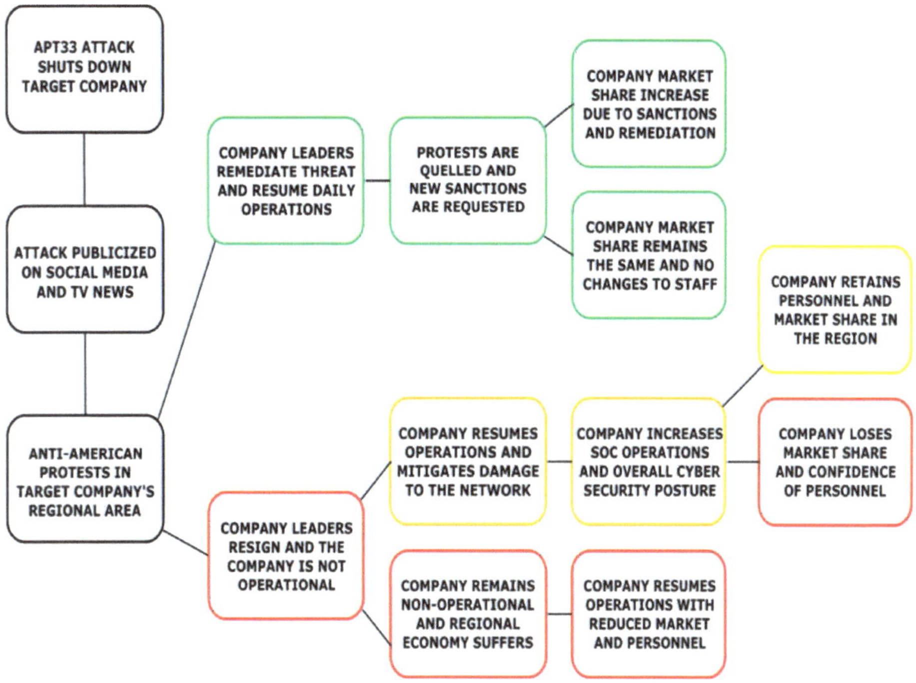

Event Trees

An *event tree* is a graphic depiction of a possible sequence of events, including junctures within the event sequence, depicting threat options and decisions.

Example Event Tree: APT33 Attack

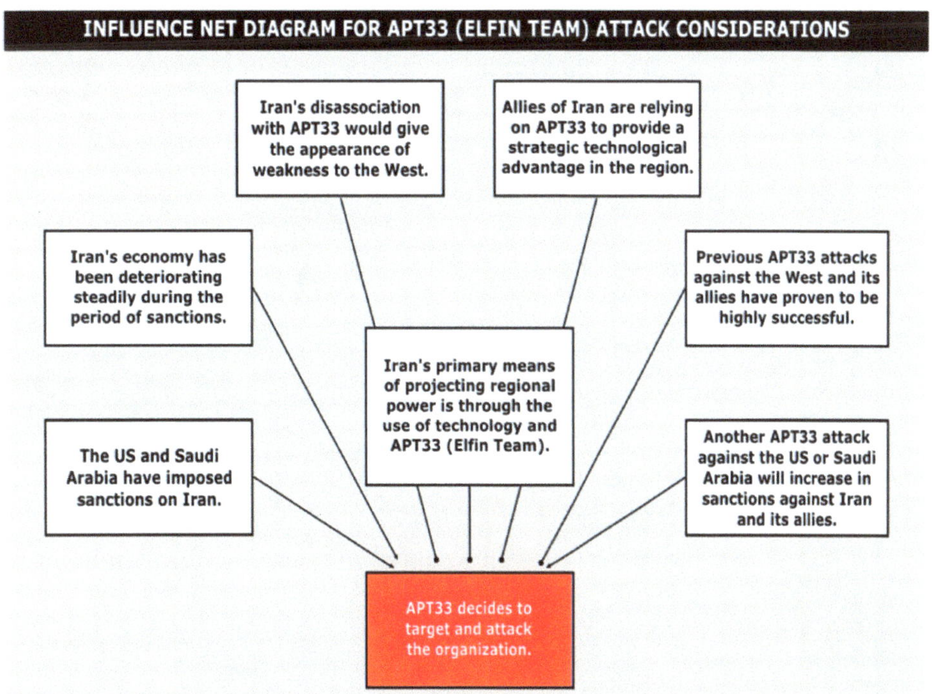

This technique is used to clarify alternative event sequences. The analyst should list multiple and mutually exclusive options that cover the spectrum of possible alternatives. It is used to depict threat options with decision points and serves to provide insight into potential vulnerabilities.

This technique assists in clarifying the presumed sequence of events and the decisions between the initial event and potential outcome. Additionally, it helps to determine collection requirements.

Subjective Probability

Subjective probability is a quantitative expression of an analyst's degree of belief in the truth of a statement relative to all other alternative possibilities by comparing the likelihood of hypotheses. It provides quantitative estimates for events, thereby mitigating the imprecision of probability phrases, and evaluates the value of additional information as it is discovered.

Once a numerical score associated with a percentage range is defined, the analyst must ensure mathematical rules are followed in each calculation. This technique also forces the consideration of alternative outcomes. Quantitative scores associated with alternatives – if properly calculated – cannot be dismissed or ignored.

Subjective Probability: Sample Table

Subjective probability does not determine a timeline for events; it attempts to allow analysts to predict an outcome by applying a percentage of probability. The sample table below visually depicts a typical use of this technique to develop numerical scores and percentage ranges.

Subjective Probability Table		
Term	Score	Range (Percentage)
Highly Probable	18-20	91 to 100
Probable	16-17	81 to 90
Highly Likely	11-15	71 to 80
Likely	8-10	61 to 70
Possible	5-7	41 to 60
Unlikely	4	31 to 40
Highly Unlikely	3	21 to 30
Improbable	2	11 to 20
Highly Improbable	1	1 to 10

Subjective Probability: Methodology (I)

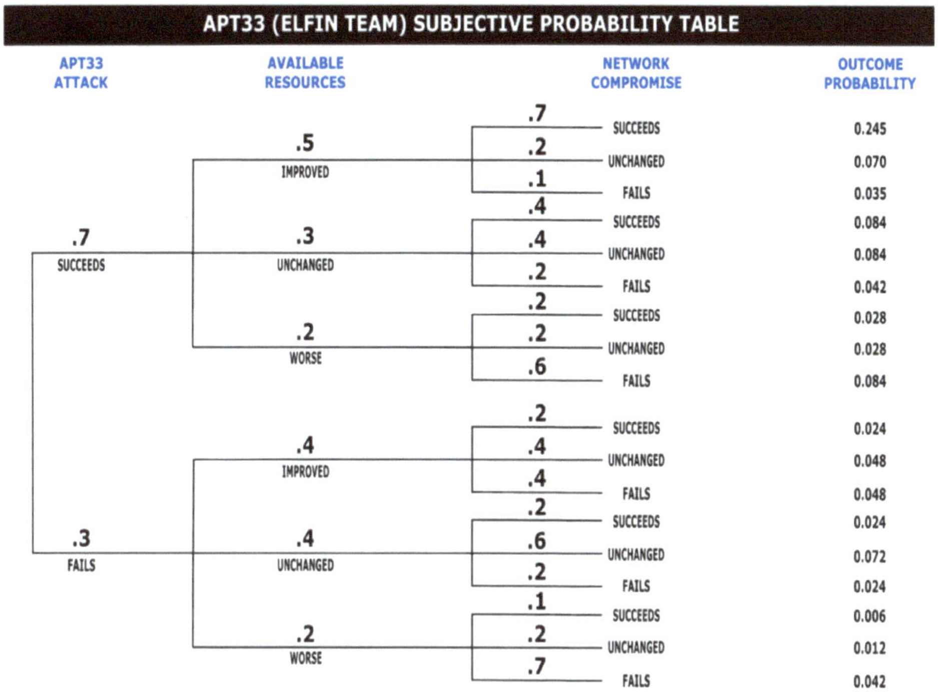

Subjective Probability: Methodology (II)

Threat Matrix: APT Comparative Analysis					
Risk Values 1-20	**Known TTPs** Documented	**APT 28** Fancy Bear	**APT 32** OceanLotus	**APT 33** Elfin Team	**APT 41** Wicked Panda
1-20	Targets Petrochemical and Energy Industries	0	0	20	1
1-20	Targets Regions of Organizational Operation	0	0	20	1
1-20	Capable of Executing Zero-Day Exploits	2	4	5	6
1-20	Uses Boot or Auto-Start Execution	15	15	10	10
1-20	Targets Non-Standard Ports and/or Protocols	15	15	10	1
1-20	Uses Encryption for C2 and Data Exfiltration	15	15	15	10
1-20	Targets Windows Servers and File Sharing Systems	15	15	15	10
1-20	Uses Spear Phishing to Compromise Accounts	20	20	20	7
Total Value 1-160		**82**	**84**	**115**	**46**

Small Number Documented	Moderate Number Documented	Significant Number Documented
Risk Values		
1-4 / **Low Probability**	5-10 / **Moderate Probability**	11-20 / **High Probability**

[THIS PAGE IS INTENTIONALLY LEFT BLANK.]

CHAPTER 10

Structured Techniques for Conducting Analysis

Knowledge Assessment Questions

NOTICE TO STUDENTS:

The following knowledge assessment questions are presented in true / false, multiple choice, and fill-in-the-blank formats. The correct answers are provided in an Answer Key at the end of Chapter 14. These questions may or may not be presented on the quizzes and/or certification exam given by the instructor of this course.

Knowledge Assessment Questions

1) __________ is an effective structured analytic technique when information elements can be broken out into categories for comparison using automated computer software.

A. Event Mapping

B. Event Trees

C. Matrices

D. Sorting

2) __________ probability does not determine a timeline for events; it attempts to allow analysts to predict an outcome by applying a percentage of probability.

A. Event Map

B. Event Tree

C. Subjective

D. Quantitative

3) An __________ is a graphic depiction of a possible sequence of events, including junctures within the event sequence, depicting threat options and decisions.

A. Event Map

B. Event Tree

C. Matrix

D. Qualitative Dataset

4) __________ are useful whenever there are more options or more intricate data than can be conceptualized at one time without a visual representation.

A. Matrices

B. Demonstration Scenario

C. Event Map

D. Event Tree

5) __________ uses a brainstorming diagram to represent the scenarios in the hypotheses linked around a central phrase representing the issue to be analyzed.

A. Extrapolation

B. Subjective Probability

C. Forecasting

D. Event Mapping

True / False

6) The three-dimensional design of matrices increases their use for collating data on complex issues and leaving out pertinent data will oversimplify issues.

A. True

B. False

7) Sorting massive amounts of data can provide insights into trends or anomalies that warrant further investigation that may have otherwise gone unnoticed.

A. True

B. False

8) The intelligence team matrix is an analytic technique used by intelligence analysts to efficiently look at information from the threat's point of view.

A. True

B. False

9) It is important to enter the information for every decision-making criteria in a column of a threat intentions matrix before moving to the next in order to mitigate hypotheses.

A. True

B. False

10) Subjective probability is a quantitative expression of an analyst's degree of belief in the truth of a statement relative to all other alternative possibilities.

A. True

B. False

CHAPTER 11

Identifying and Exploiting Network Relationships

Lecture Summary

KEY KNOWLEDGE POINTS

Relationship Modeling and Analysis

Network Model Types

Overview of Network Analysis

Centrality and Equivalence

Threat Network Analysis

Target Network Analysis

Relationship Modeling and Analysis

Relationships among entities – people, places, things, and events – are perhaps the most common and frequently used subject of intelligence modeling. This type of model takes a significant amount of time to create. The model must be constantly maintained, and that continuous maintenance demands effort on the part of analytical teams. The constant inputs needed to ensure the model remains current is referred to as "*feeding the beast*."

These models are highly effective in analyzing modern issues, using sophisticated software and powerful graphical displays. This level of presentation is also effective for persuading customers to accept analytical results.

There are four levels of relationship models, each using increasingly sophisticated analytic approaches: *hierarchy*, *matrix*, *link*, and *network models*. Listed below is a brief description of each; they will each be addressed in greater detail later in the domain.

1. *Hierarchy Models*: These are typically represented as simple tree models.

2. *Matrix Models*: These models contain numerous grids which reveal relationships.

3. *Link Models*: These models are complex tree structures that show relationships between nodes.

4. *Network Models*: These models contain multiple hierarchies that include *social networks* (human relationships) and *target networks* (entity relationships).

The *Netwar* Model

The *netwar* model of multidimensional conflict (developed by RAND Corporation) between opposing networks is increasing applicable to all intelligence and the network analysis tool of choice. It consists of the four primary components listed below.

1. **Network**: Also referred to as *system*, this contains groups forming a unified whole.

2. **Node**: This represents any element of a system.

3. **Cell**: This represents nodes which are formed around a specific process.

4. **Link**: This represents physical or functional relationships between nodes.

Link Models

Link modeling is routinely used in cybersecurity and law enforcement to identify Advanced Persistent Threats, Dark Web criminals, and terrorist organizations. This type of model is used to display relationships among people, and relationships between people and events. It is useful for identifying complex relationships as human minds cannot assimilate and organize all collected data.

Used primarily in operational and strategic planning, the goal of link analysis is to develop precise and valid inferences. Software has contributed significantly to this process in recent years given its ability to analyze and display collected data from a variety of different perspectives.

Network Models

Network modeling and analysis has proven to be highly effective in intelligence organizations that focus primarily on both friendly and adversarial networks. It has become a critical tool for intelligence organizations. This technique is useful for target modeling and planning offensive tactics and strategies (the targeting of threats).

Relationships in network models can be either positive or negative, with the link thickness used to indicate the strength of relationships. It is also a powerful tool used in *causal modeling* and useful for the creation of *influence nets*. To aid the analyst using this type of model, color coding can be used to quickly identify the types of nodes being represented in the model.

Network Model Types

There are numerous types of networks but for practical purposes, they are all target networks, even though the network is not necessarily hostile. In many cases, the networks are friendly and allied entities. In others, the networks include neutral entities targeted to be influenced – either to remain neutral or be converted to allies.

Network models may include several network types such as threat, social, financial, or organizational and tend to be characterized by the nature of their relationships. A *target network* can be any combination of types as well, and it is common for the network to include more than one form or type. A few common target network types are listed below.

1. **Functional Networks**: These networks are formed for a specific purpose and their activities are based on the member's skills and expertise.

2. **Family and Cultural Networks**: These networks include members bonded by culture, language, religion, ideology, or proximity of geographic ties.

3. **Virtual Networks**: These networks include actors that seldom (if ever) physically meet.

Manual Modeling

When network modeling was first developed as an analytic methodology, the target models were created manually drawing circles and links on paper. The steps listed below will guide analysts engaged in the manual modeling of networks.

1. *Understand the environment*: Consider all six *PMESII* operational variables and *ASCOPE* environmental factors to assist in providing context.

2. *Select or create a network template*: Use foundational analytic methods to conduct pattern, link, or *social network analysis* (SNA).

3. *Populate the network template*: Use the "snowballing" technique to develop model content if needed.

Computer-Assisted and Automated Modeling

Although manual modeling is still used, commercially available network tools and software are available to help make the analyst's efforts more efficient. Analysts should strive to automate the process completely, creating the models directly from raw data that is collected.

These network models are prevalent and typically use data mining and text-mining techniques to gather raw data. To increase the probability of accurate models, analysts must ensure to evaluate the sources of data and validate the results. It is common for massive datasets to generate a significant number of false alarms.

Network Analysis

Network analysis involves addressing classic intelligence questions – *what, where, when, who, how, and why* – and placing the answers in a format the customer can understand and act upon (actionable intelligence). This type of analysis helps identify both the function and importance of network entities. The analysis methods listed below answer specific intelligence questions.

1. **Network Pattern Analysis**: Identifies *what*, *where*, and *when* components of the intelligence question.

2. **Social Network Analysis**: Identifies the *who* of the intelligence question.

3. **Network Nodal Analysis**: Identifies *how* and *why* components of the intelligence question.

Social Network Analysis (SNA)

When intelligence analysts refer to network analysis, they are often referring to *social network analysis* (SNA) using platforms such as Twitter and Facebook. It is used to identify and assess relationships. Links – represented visually or mathematically – reveal nodal relationships. This type of analysis identifies key individuals, referred to as *actors*, within the network. Two fundamental analytic concepts are of crucial importance in this analysis: *centrality* and *equivalence*.

Centrality

Centrality refers to the sources and distribution of power in a social structure, and the network perspective suggests power is derived from relationships. This term applies to actors that occupy an advantageous position within the network. The network position can reveal their constraints and their ability to constrain others. It is important for the analyst to assess the actor's location if possible. To gain maximum insight into the various roles and groupings within the network, the analyst should answer the questions listed below.

1. *Who are the leaders and connectors?*

2. *Where are the clusters and what actors are residing within them?*

3. *Who is in the core of the network?*

Equivalence

The disruptive effectiveness of removing one individual or a group of individuals from a network depends not only on centrality but on their uniqueness as well. *Equivalence* is useful for strategic targeting missions and is closely related to the concept of centrality. Nodes in the network tend to have special roles and may be unique (they have no equivalent actors). These actors are extremely difficult to replace when lost. All valuable targets have centrality; the most valuable targets have no equivalents.

Target Value Analysis (TVA)

Target Value Analysis (TVA) is a method of identifying and prioritizing *High-Value Targets* (HVTs) and *High-Payoff Targets* (HPTs) to determine which should be targeted and engaged. Both types of targets have a high degree of network centrality.

1. **High-Value Targets (HVTs)**: These individuals are important to the adversary's leadership and removing them is important to the mission's success.

2. **High-Payoff Targets (HPTs)**: These individuals are critical HVTs. The loss of these individuals will degrade the adversary's potential for success and directly support the friendly mission's success.

Developing a High-Payoff Target (HPT) List

Engaging targets that the adversary can least afford to lose deprives them of their functionality and forces them to alter their established decision cycle. The steps listed below are used by intelligence analysts to create HPT lists.

1. Identify and prioritize all HVTs in the target network.

2. Eliminate the HVTs currently beyond reasonable reach from the list.

3. Eliminate the HVTs beyond current targeting abilities from the list.

4. Eliminate HVTs beyond reasonable timelines from the list.

5. The remaining HVTs now become the HPT list for the analyst.

6. Coordinate relevant Information Requirement (IR) taskings for each HPT.

7. Disseminate the list to collection teams and all authorized personnel involved.

Network Nodal Analysis

Nodes in a target network can include people, places, objects, and organizations, all of which can be treated as separate networks if required. This analysis may involve assessing the performance of objects such as datacenters or malware. Cyber threats can be large, complex, and identifying critical objects for analysis may prove challenging. An analyst attempting to identify the most critical network nodes must be able to accomplish the tasks listed below.

1. Conduct analysis of each individual node.

2. Understand the behavior of each of the involved nodes.

3. Have insight into the activities taking place within each of the nodes.

Threat Network Analysis

Threat networks have been defined as being "comprised of people, processes, and places – components that can be identified, targeted, and exploited." They are opposed to friendly networks, tend to have state backing, and can assume many forms and functions. Examples of threat networks include insurgents, violent jihadists, extremist groups, and international criminal organizations.

A premise of threat network modeling is that these networks have vulnerabilities that can be exploited once they are identified by intelligence analysts. To be successful in this task, analysts must understand how the network operates and identify actions to exploit potential vulnerabilities. These vulnerabilities will vary based on the network location: some networks are territorial, and some are transnational.

Larger and more sophisticated threat networks have access to *PMESII* resources (Political, Military, Economic, Social, Information, and Infrastructure), making them more challenging targets. Many are connected to social structures and use global information networks for their recruiting, funding, and operational needs.

Analysts develop a detailed understanding of how threat networks function by identifying its elements, processes, and how the components interact. Listed below are several critical components common to threat networks that should be identified and understood by the analyst.

1. **C2 Structure**: The command-and-control structure of the organization.
2. **Closeness**: Commonalities internal and external to the organization.
3. **Expertise**: The knowledge, skills, and abilities of the actors within the organization.
4. **Resources**: The ability to secure financial and public support.
5. **Adaptability**: The ability to learn and adjust their behaviors as needed.
6. **Sanctuary**: Access to safe areas to conduct missions without interference.

Some threat networks must compete for resources, support, and territories, which requires communication that can be exposed and mapped by linking. These networks tend to be organized as cells and may have several strengths and exploitable weaknesses. Listed below are several strengths and weaknesses an analyst should investigate when analyzing targeted threat networks.

1. *Strength*: The ability to adapt over time
2. *Strength*: The ability to blend into the local population
3. *Strength*: The ability to quickly replace losses of key personnel
4. *Strength*: The ability to quickly recruit new members
5. *Weakness*: The tendency to follow standard operating procedures (SOPs) from past successes
6. *Weakness*: Isolation for security purpose restricts communication and evolution
7. *Weakness*: Lack of evolution in methodologies makes the target predictable

Target Network Analysis

Analysts usually apply an extension of social network analysis to all intelligence work, and SNA techniques work for almost all *target network analysis* missions.

Entities observed in *social network analysis* (SNA) tend to be people, but within the scope of target network analysis the focus can be anything including associated locations and nodal activities. The analyst's goal is to reveal patterns in the network and identify future targets. Listed below are the three common types of networks an analyst will encounter during the intelligence analysis.

1. Threat Networks
2. Neutral Networks
3. Allied Networks

Automating Network Analysis

Target network analysis has become one of the principal tools for dealing with complex systems because of computer-based analytic methods and software. One of the most popular and well-established tools created for this type of analysis is the *Organization Risk Analyzer* (ORA). Listed below are popular features and capabilities associated with this software.

1. The ability to group network nodes.
2. The ability to identify patterns of network significance.
3. The ability to identify key network actors and groups.
4. The ability to identify potential vulnerabilities.
5. The ability to model spatial and temporal changes.
6. The ability to employ sophisticated graphical techniques
7. The ability to store and manipulate massive datasets.

NETANOMICS

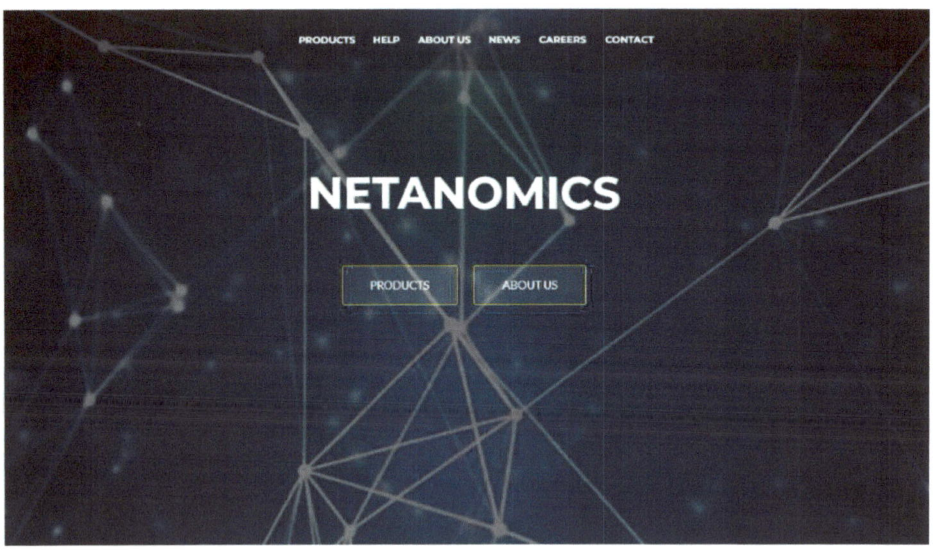

netanomics.com

CHAPTER 11

Identifying and Exploiting Network Relationships

Knowledge Assessment Questions

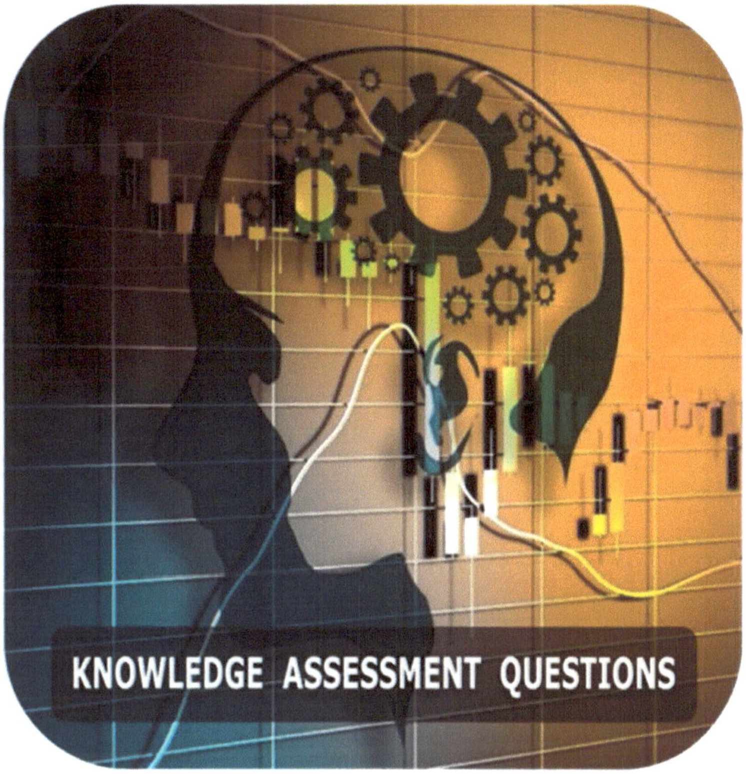

NOTICE TO STUDENTS:

The following knowledge assessment questions are presented in true / false, multiple choice, and fill-in-the-blank formats. The correct answers are provided in an Answer Key at the end of Chapter 14. These questions may or may not be presented on the quizzes and/or certification exam given by the instructor of this course.

Knowledge Assessment Questions

1) There are four levels of relationship models, each using increasing sophisticated analytic approaches: hierarchy, link, __________, and network models.

A. Host

B. Node

C. Cell

D. Matrix

2) __________ modeling is routinely used in cybersecurity and law enforcement to identify Advanced Persistent Threats, Dark Web criminals, and terrorist organizations.

A. Matrix

B. Node

C. Link

D. Cell

3) Target network analysis has become one of the principal tools for dealing with complex systems as a result of __________-based analytic methods and software.

A. Relationship

B. Computer

C. Analysis

D. Threat

4) __________ in a target network can include people, places, objects, and organizations, all of which can be treated as separate networks if required.

A. Nodes

B. Links

C. Edges

D. Domains

5) __________ Analysis is a method of identifying and prioritizing High-Value Targets (HVTs) and High-Payoff Targets (HPTs) to determine which should be engaged.

A. Matrix

B. Computational

C. Network

D. Target Value

True / False

6) The network model of multidimensional conflict between opposing networks is increasingly applicable to all intelligence and the network analysis tool of choice.

A. True

B. False

7) There are numerous types of networks but for practical purposes, they are all target networks, even though the network is not necessarily hostile.

A. True

B. False

8) A premise of threat network modeling is that these networks have strengths that can be exploited once they are identified by intelligence analysts.

A. True

B. False

9) Engaging targets that the adversary can least afford to lose deprives them of their functionality and forces them to alter their established funding cycle.

A. True

B. False

10) Equivalence refers to the sources and distribution of power in a social structure, and the network perspective suggests power is derived from relationships.

A. True

B. False

CHAPTER 12

Using Simulations for Intelligence Decisions

Lecture Summary

KEY KNOWLEDGE POINTS

Overview of Simulation Modeling

Types of Simulations

Creating and Running a Simulation

Simulations Used in Intelligence Analysis

Decision Modeling and Simulation

Game Theory and Decision Response

Chapter 12: *Using Simulations for Intelligence Decisions*

Overview of Simulation Modeling

Simulation models are mathematical descriptions of the interrelationships believed to determine a system's behavior under different conditions. It is impractical for analysts to measure all required variables used to estimate real-world performance. Intelligence estimates made under different conditions help customers to choose among alternatives.

Simulations determine the likely outcomes of actions with multiple variables. They are useful for making long-range forecasts as exact numerical estimates are not needed for the model.

Types of Simulations

1. **Deterministic**: These models are created with a spreadsheet consisting of rows and cells. The inputs are fixed by cell numbers and only one solution is presented.

2. **Stochastic**: These models are used when uncertainty exists about proper inputs. Values are assigned to uncertain inputs (as opposed to fixed cell numbers) and the model considers all possible outcomes. This is also referred to as a *Monte Carlo* simulation. The challenge of a stochastic simulation is to select the right uncertainties for the input variables which represent the correct probability distribution.

The simulation models typically used by intelligence analysts are classified as *continuous simulations*, *discrete-event simulations*, or a *hybrid* of the two.

3. **Continuous Simulations**: These models use the same type of inputs but assume the system in question changes with time.

4. **Discrete-Event Simulations**: These models use the same type of inputs but assume the system changes states as events occur.

5. **Hybrid Simulations**: As the name implies, these models are a combination of continuous and discrete-event simulations.

Creating and Running a Simulation

Sophisticated simulation models can be complicated, and independent checks against real-world results are difficult to run on them. Listed below are the general steps analysts follow to create simulation models.

1. *Formulate the issue*. The analytic results as referred to as "*Measures of Effectiveness* (MOEs)" of "*Figures of Merit*."

2. *Identify the information needed for input*.

3. *Select the appropriate simulation software*.

4. *Validate the model*. Proven and validated models are preferred by customers.

5. *Run the simulation and interpret the results*. Continue this process as needed.

Chapter 12: *Using Simulations for Intelligence Decisions*

Simulations Used in Intelligence Analysis

Intelligence analysts use interactive simulation models extensively to assess the performance of adversarial threats, capabilities, and even financial economies. Listed below are six of the most prevalent simulation models that analysts are most likely to encounter.

1. Network Simulations
2. Offensive Capability Simulations
3. Weapons System Simulations
4. Economic Simulations
5. Geospatial Simulations
6. Political and Social Simulations

Network Simulations

Network simulations are a powerful tool used by intelligence analysts for modeling and analyzing threat, neutral, and allied networks. These models demonstrate network flows of variables such as money, information, technology, and disease. This type of modeling is also efficient as performing the tasks listed below.

1. Assessing the impact on a network by removing nodes.
2. Assessing the most important nodes and links in a network.
3. Assessing estimated network recovery times.

Offensive Capability Simulations

Also referred to as "war gaming," *offensive capability simulations* are closely related to offensive cybersecurity simulations and often incorporate them. They are used primarily by defense organizations for planning systems acquisition, determining force mixes, and personnel training (both technical and combat). These types of simulations incorporate several types of sub-models, the three most pervasive being listed below.

1. **Vulnerability and Cyber Effect Models**
2. **System Reliability Models**
3. **Force-on-Force Campaign Models**

Weapons System Simulations

Weapons systems simulations range in complexity from simulating performance of a single entity to that of elaborate interconnected systems. They require high-speed supercomputers capable of processing complex simulation codes. Many commercial simulation models already exist and are available for the civilian market. They address a wide range of system simulations such as radar and communication systems, satellite orbits, aircraft flight plans, and shipping routes. The risk posed to analysts is that details generated by these simulations can overwhelm analysts, causing them to lose sight of their primary objectives.

Economic Simulations

Economic models (also referred to as *econometric models*) are used to simulate the performance of an entire economy or any segment of it such as competitive industries. These models provide a quantitative description of the system. They incorporate several possible hypotheses to describe how target economic systems function.

They are used extensively by intelligence and financial organizations to assess trade, balance of payments, energy production, distribution, and even support diplomatic negotiations. Analysts should be aware that customers tend to prefer econometric models.

Geospatial Simulations

One of the most powerful applications of *geospatial simulations* comes from combining them with models such as *Operational Requirements Analysis* (ORA). Simulations can take many forms and have many applications, such as determining locations for pandemic treatment centers, terrain mapping, and developing flood models.

They rely on radar and *synthetic aperture radar* (SAR) and require the analyst to input geographic coordinates for nodes to generate accurate outcomes. Geospatial and network simulations are often combined to predict population movements.

Political and Social Simulations

Political and social simulations have been developed to analyze the interactions of political parties, diplomatic initiatives, and clashes across cultural fault lines. The simulations receive data from many sources such as historical patterns, web activity, and social media inputs. Listed below are several common uses for this type of analysis.

1. Warn of possible **natural disasters**
2. Warn of possible **disease activity**
3. Warn of possible **trends in crime**
4. Warn of possible **geopolitical events**

Decision Modeling and Simulation

Decision modeling and simulation analyzes the personality, problem-solving styles, values, goals, behaviors, and environments of individual leaders. These types of models are primarily based on the six types of models listed below.

1. Rational Models
2. Administrative Models

3. Cultural Models

4. Emotional Models

5. Collective Decision-Making Models

6. Game Theory

Rational Models

Rational decision-making requires the systematic evaluation of costs or benefits accruing to courses of action that may be taken by an individual. The process begins with the analyst identifying all possible choices involved. Values – referred to as a "Cost-Benefit Analysis" – are assigned for the possible outcomes. The probability of potential outcomes is then expressed in the model. The *rational model* is based on the two foundational concepts listed below.

1. *Optimal Choice*

2. *Expected Utility Theory* (EUT)

Administrative Models

The tendency of leaders to adopt suboptimal choices leads to a variant of the rational decision-making simulation referred to as the *administrative model*. The model starts with an assumption that the effort to find an optimum decision has not yet been made by the leader – the complexity of a given problem dissuades efforts.

That leads to a tendency to select several possible outcomes that are not "best case scenarios" but meet the test of "good enough." This model disregards the foundation of *Expected Utility Theory* (EUT) and instead makes the assumptions listed below.

1. People tend to have incomplete information.

2. People have time pressures which impact their decisions.

3. People are always beset with conflicting preferences.

Cultural Models

Behavior cannot be predicted with any degree of analytical confidence without the analyst putting it in the decision maker's social and cultural context. Analysts working with *cultural models* must understand the elements of the culture, such as how youths are trained for adult lives and how that training defines the individual's importance in life.

Culture defines the ethical norms of the collective. It dictates values and constraints decisions (it is a constraining social and environmental force). The two risks posed to analysts in this environment are listed below.

1. *The irrational behavior trap*. Analysts may be inclined to view the actions of others as irrational when compared to the norms and behaviors of their own culture.

2. *The mirror imaging trap*. Discussed earlier in the course, mirror imaging occurs when analysts impose their own norms and beliefs on targets that do not share them. This typically presents itself in the question, "*what would I do in a similar situation*?"

Emotional Models

An important aspect of the decision process to consider when analyzing behavior is the emotional one, a dominant component in many influential leaders. The *emotional model* was designed to take this reality into account. Emotional aspects must be addressed. Exciting and challenging things tend to attract humans. Pride and revenge are also motivators.

Established methods for personality profiling are available to analysts but they must also the consider two realities of the human condition listed below.

1. Personal idiosyncrasies and emotional predisposition tend to prevail.

2. Calculated assessments are secondary to those who base decisions on emotional predisposition.

For example, given the same set of circumstances, a leader who spent his college career as a heavyweight boxing champion will most likely have a different predisposition than a leader who served as the president of the college chess club. Profile models are still effective in assessing likely decisions and outcomes if the analyst is aware of the distinctions that drive these idiosyncrasies and predispositions.

Collective Decision-Making Models

Decisions are often made by a group and such cases require that intelligence analysts employ a *collective decision-making model* for their predictions. Although groups have more inherent complexity than individuals, it is often easier to predict the actions of groups. Listed below are three dominant features of this model that must be considered by analysts.

1. *Power is decentralized*. Profiling one individual assumed to be the organizational leader will not provide all the possible and likely outcomes to the analyst.

2. *Multiple competing goals exist*. The number of goals – desired outcomes to varying scenarios – far exceeds those of a single leader. Competing goals exist between individuals in the collective and within the individuals themselves.

3. *All decision result from bargaining*. Individuals bargain with individuals, who then form coalitions that bargain with each other. The success of the bargaining initiative is contingent upon the negotiation skills of the individual and the ability to form consensus within a coalition.

This is a complex process of conflict resolution, negotiation, building consensus, and force of will. All decisions (outcome scenarios) will be the product of compromises and typically do not match the initial desired outcome. A benefit to the analyst in this scenario is historical precedent. The actors, methods, and outcomes associated with collective decision-making tend to follow predictable behaviors that are based on previous successes.

Game Theory

Game theory analyzes the decision process of two or more parties in conflict or cooperation and is valuable in simulation for its use of mathematical equations.

The game (simulation model) assumes two or more players. Each player is required to determine how the other player will respond, then determine their likely response to the other player's choice. The two goals listed below share the requirements of the players participating in the game.

1. Each player must estimate the move of the other players.

2. Each player must determine how their initial decision will impact the end of the game.

The premise of the exercise is simple and straightforward. Each player analyzes all the information available and then identifies an initially preferred decision. The difficulty lies – both for the players and the analyst – in understanding the decision processes of others involved in the game.

CHAPTER 12

Using Simulations for Intelligence Decisions

Knowledge Assessment Questions

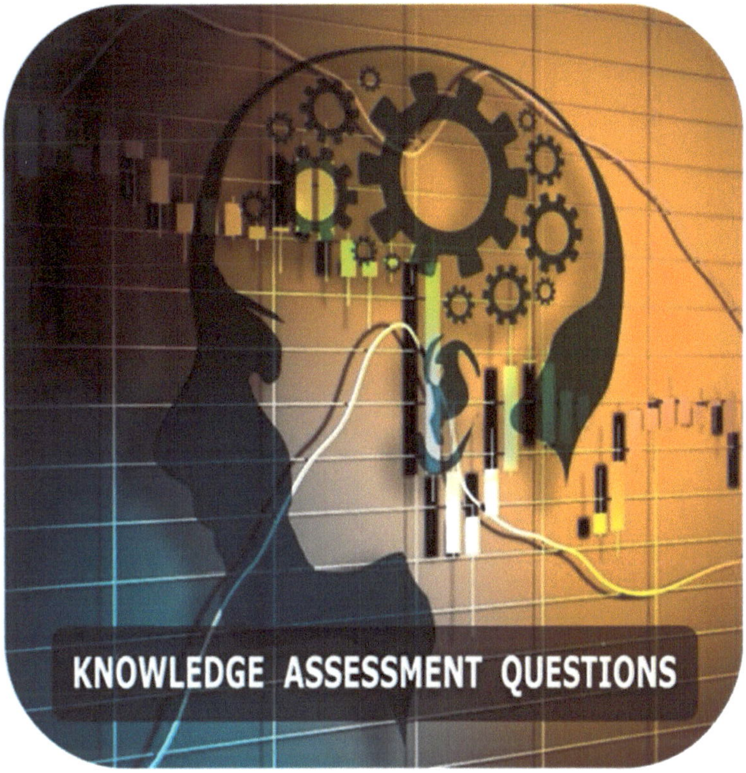

NOTICE TO STUDENTS:

The following knowledge assessment questions are presented in true / false, multiple choice, and fill-in-the-blank formats. The correct answers are provided in an Answer Key at the end of Chapter 14. These questions may or may not be presented on the quizzes and/or certification exam given by the instructor of this course.

Knowledge Assessment Questions

1) The challenge of a __________ simulation is to select the right uncertainties for the input variables which represent the correct probability distribution.

A. Stochastic

B. Deterministic

C. Continuous

D. Hybrid

2) Intelligence analysts use __________ simulation models extensively to assess the performance of adversarial threats, capabilities, and even financial economies.

A. Snapshot

B. Interactive

C. Static

D. Intelligence

3) __________ analyze the decision process of two or more parties in conflict or cooperation and is valuable in simulation for its use of mathematical equations.

A. Collective Decision-Making Models

B. Emotional Models

C. Game Theory Models

D. Cultural Models

4) The tendency of leaders to adopt suboptimal choices leads to a variant of the rational decision-making simulation referred to as the __________ model.

A. Cultural

B. Emotional

C. Rational

D. Administrative

5) __________ models are used to simulate the performance of an entire economy or any segment of it such as competitive industries.

A. Economic

B. Geospatial

C. Political

D. Social

True / False

6) The simulation models typically used by intelligence analysts are classified as continuous simulations, discrete-event simulations, or a hybrid of the two.

A. True

B. False

7) Composite simulations are a powerful tool used by intelligence analysts for modeling and analyzing threat, neutral, and allied networks.

A. True

B. False

8) An important aspect of the decision process to consider when analyzing behavior is the administrative one, a dominant component in many influential leaders.

A. True

B. False

9) Rational decision-making requires the systematic evaluation of costs or benefits accruing to courses of action that may be taken by an individual.

A. True

B. False

10) One of the most powerful applications of geospatial simulations comes from combining them with models such as *Operational Requirements Analysis* (ORA).

A. True

B. False

CHAPTER 13

Making Decisions Based on Potential Outcomes

Lecture Summary

KEY KNOWLEDGE POINTS

The Purpose of Outcome Scenarios

The Role of Outcome Scenarios

Types of Scenarios

Scenario Perspectives

How to Construct Scenarios

Using Red Team Analysis

Chapter 13: *Making Decisions Based on Potential Outcomes*

The Purpose of Outcome Scenarios

The purpose of *outcome scenarios* is to highlight significant forces that may change the future, making forces visible and easier for an analyst to recognize. These scenarios help all involved to anticipate the future and craft better responses to subsequent events by effectively communicating non-existent situations.

They convey the results of decisions to action that are difficult to quantify in complex situations. The scenarios provide tactical, operational, and strategic benefits as they present a set of alternative scenarios that reflect a range of plausible outcomes.

In an alternative future depicted by a scenario, a customer should be able to make decisions and develop strategies using outcomes developed by the analyst. Predicting the future in detail is not possible – details are controlled by divergent factors while trends are controlled by convergent factors. This process allows for "most likely" outcomes defined by the indicators which reveal them. The analyst must track the indicators for specific scenarios to form hypotheses for these outcomes.

The Role of Outcome Scenarios

Outcome scenarios present several scenarios for an uncertain future about what potential events may lie ahead, each one modeling a plausible outcome. In short, they convey possible courses of events. Some events may be difficult for the customer to grasp as they may be far removed from anything originally envisioned.

All anticipatory analysis results in scenarios – typically presented to the customer as a story – that describe the future state of the target. These results, used for planning and decision making, tend to be presented with the assistance of augmented reality technologies.

Types of Scenarios

Analysts use the four basic types of outcome scenarios listed below.

1. *Demonstration Scenarios*
2. *Driving-Force Scenarios*
3. *System-Change Scenarios*
4. *Slice-of-Time Scenarios*

Demonstration Scenarios

When analysts create a *demonstration scenario*, they identify the high-impact outcomes and then the plausible paths that may lead to that result. These are a high-impact / low-probability product that identify triggering events along the scenario path. The triggering events represent decision or occurrence points that may shape future outcomes.

The analyst should identify later occurrences of events on these paths which may indicate the development of a specific scenario. Attention must also be focused on the triggering events as they may change the high-impact outcome originally predicted.

Driving-Force Scenarios

A *driving-force scenario* is an estimative approach for projection in which major forces acting on a target are examined to determine how they are changing. Analysts should create multiple driving-force scenarios and identify key uncertain forces and factors, as these key forces will tend to shape potential outcomes. Alternative change levels should be identified for each as well.

The analyst will develop a matrix or tree to help the customer visualize results and shows outcomes for the assumed changes. This visualization can describe the effects of changing forces and alternative force levels that remain constant.

System-Change Scenarios

A *system-change scenario* is a variant of the driving force scenario and accounts for the fact that alternative force levels remain constant in the estimative model. This process considers all *PMESII* variables, accounts for changes in existing forces, and accounts for the introduction of new force on the target.

The scenario relies on *cross-impact analysis* that identify interactions among events and interactions that serve to develop alternative outcomes. This tends to be an especially difficult scenario for analysts to develop as it requires a focused examination (and understanding) of the interrelationships of forces.

Slice-of-Time Scenarios

A *slice-of-time scenario* starts at a time in the future in which a synthesis of explicit or assumed forces has already created a specific outcome. It resembles the demonstration scenario in that an endpoint outcome is described but not specific pathway to the endpoint is provided. It is not typically used in actionable intelligence; there is no focus on the forces leading to the outcome and it provides minimal indicators for the analyst to monitor. Results of these scenarios are often seen in policy papers that attempt to define a specific environment.

Scenario Perspectives

Scenarios are models, and like the models utilized by intelligence analysts which were discussed in previous domains, they can be either *descriptive* or *normative*. The scenario usually begins in the present and unfolds to a future time. Extrapolations are *momentum scenarios* – they are dominated by *inertia* and assume no opposing forces arise to change trends – and tend to be a poor choice for long-term assessments. Customers prefer *normative scenarios* (predictive) as their goal tends to be to develop a course of action.

How to Construct Scenarios

Scenario planning is a standardized and systematic methodology that derives from well-known modeling, simulation, and gaming theories used in the past. The six steps listed below are common to the process analysts use to construct general scenarios and are explained in detail in this domain.

1. Define the issue and the target framework.
2. Identify and rank the driving forces.
3. Select the appropriate scenario logic.

4. Flesh out the scenario.

5. Draw out the implications.

6. Identify and monitor indicators.

Step 1: Define the Issue and Target Framework

For most intelligence issues, there are typically several customers interested in receiving intelligence who have different interests or questions to be answered. The steps listed below will assist the analyst in accomplishing this task.

1. Ask relevant preliminary questions.

2. Define the issue or the problem to be solved.

3. Conduct a key assumptions check.

4. Select the relevant issue categories.

5. Decompose the issue and sub-issues.

6. Select the proper type of target model to be used.

7. Create the target framework for the issue.

8. Develop alternative target frameworks as needed.

Step 2: Identify and Rank the Driving Forces

Because scenarios are a way of understanding the dynamics driving the future, analysts attempt to identify the primary driving forces shaping the present. The steps listed below will assist the analyst in accomplishing this task.

1. Isolate the known driving forces.

2. Identify specific forces critical to the outcome (some driving forces impact everyone).

3. Identify any asymmetric forces which may be present.

4. Rank the driving forces by **importance**.

5. Rank the driving forces by **uncertainty**.

Step 3: Select the Scenario Logic

Analysts want to finish with a few scenarios – typically four or fewer – that provide their customer with the basis for making the best decision possible. This is probably the most important step as it is derived from differences in the model.

1. Identify and work with the known issues, targets, and forces.

2. Reshape and reframe the components as needed.
3. Draw out the less obvious elements of the various components.
4. Review the components for emerging patterns.
5. Note that a few forces have the potential to different the outcome. These forces are referred to as "*Scenario Drivers*."

Step 4: Flesh Out the Scenario

Analysts now need to return to step two of this process, "*Identify and Rank the Driving Forces*," and use those elements to further flesh out the scenario. This allows analysts to expand the detail of scenario elements. The three techniques listed below are commonly used by analysts to accomplish this mission.

1. **Case-Based Models**
2. **Contextual Mapping**
3. **Cross-Impact Analysis**

Case-Based Models

Case-based models are the foundation for a type of analysis called case-based reasoning which uses reasoning by analogy or reasoning by history. This process uses previous experiences to solve new issues. Analysts consider a previous situation similar to the current one and use the previous situation to help solve the new issue. Of course, previous situations must be adapted to meet new demands. Analytical reasoning formed from precedent is used to interpret new facts. Reusing knowledge – duplicating the model – is a useful tool for analysts when faced with similar situations.

Contextual Mapping

The technique of *contextual mapping* is used to identify plausible sequences of development in a field and relate them to developments in a different field. It is useful for forcing fresh perspectives and predicting actual developments. This process requires experienced analytical experts that are familiar with this method and the topic of inquiry. It is primarily used in technical forecasting in which future developments are contingent upon external forces. It typically takes the form of a graphical display with timelines and presented to customers to show how interconnecting paths lead to developments.

Cross-Impact Analysis

Cross-impact analysis specifies how one event will influence the timing, model of impact, and likelihood of another event in a different but associated field. It is used to support *system-change scenarios* by showing interactions among developments. This type of analysis is used extensively to accomplish the two missions listed below.

1. Model the interaction of future **events**.
2. Model the interaction of future **trends**.

Step 5: Draw Out the Implications

Having built out a set of potential future scenarios and outcomes, the analyst must examine each of them carefully in the context of the original question. Almost every scenario will have tactical, operational, and strategic impacts. When these impacts are unforeseen, they are referred to as "*Laws of Unintended Consequence*." Everyone involved in the process – including the primary customers of intelligence, policy executives, and decision-makers – must consider all possible outcomes. Once that reality is embraced, options for the specific outcomes must be developed accordingly.

Step 6: Identify and Monitor Indicators

The job of the intelligence analyst (and customer) becomes one of *identifying and monitoring indicators* once the scenarios have been completed. Analysts must look for the leading indicators to identify the scenario (or combination of scenarios) taking place. This observation can also indicate the scenarios that are not taking place. Monitoring may also involve watching trends. Analysts should select options to focus customer decisions and identify those potential outcomes which will require preparation to address. In the discipline of intelligence, the outcomes requiring preparation are referred to as "*Indicators of Signposts of Change*."

Devil's Advocates

On an intelligence team, a devil's advocate is someone who takes a position he or she does not necessarily agree with to provoke a debate. The analyst assuming this role seeks to directly engage others in an argumentative discussion process. This individual will attempt to test the validity and the quality of the intelligence product. The process involves pre-event preparation, effort, and tends to accomplish the goals listed below.

1. Identify weaknesses in the product's **logic**.
2. Identify weaknesses in the product's **assumptions**.
3. Identify weaknesses in the product's **conclusions**.

Overview of Red Team Analysis

Red Teams are a form of challenge analysis used by intelligence teams that adopt the mindset and mentality of the adversary to become the "enemy." This process is used to criticize intelligence assessments employing knowledge of the adversary's culture and practices. This requires *subject matter expertise* (SME) knowledge of the adversary.

Red Team analysis is **not** role-playing as the opposing team is not submerged in the adversarial role. This process can help detect errors in measurements and provides an overall reduction of cognitive biases. Red Teams approach this endeavor to address the mindsets of how the adversary may think, address the beliefs of what the adversary may need, and address the means the adversary may use to achieve their objectives. It is used at the end of the assessment process to address a diverse set of scenario contexts.

Analysts using this process must believe in the Red Team's effectiveness and be willing to accept the findings. Leadership must also support this style of challenge analysis. It has been historically difficult for analysts to accept the Red Team findings of challenge analysis and for leaders to afford teams the time to fix problems that Red Team's may identify and reveal.

Red Teams: Understanding Adversaries

Red teaming is often used to analyze an adversary's decision-making, assessing likely strategies and tactics, and predicting the outcomes of thought processes. The four considerations listed below must be acknowledged to understand the adversary and for this process to be effective.

1. **Different Beliefs**: Adversaries tend to have a radically different mental model of the world.

2. **Different Objectives**: Adversaries may engage in actions for the purposes of creating instability, attention, and engaging in ideological expansion.

3. **Different Capabilities**: Adversaries can increase their capabilities and effectiveness when they have les concern about internal opposition.

4. **Different Moral Constraints**: When considering adversarial objectives versus moral cost and human rights, it is not surprising to discover that objectives take precedence.

CHAPTER 13

Making Decisions Based on Potential Outcomes

Knowledge Assessment Questions

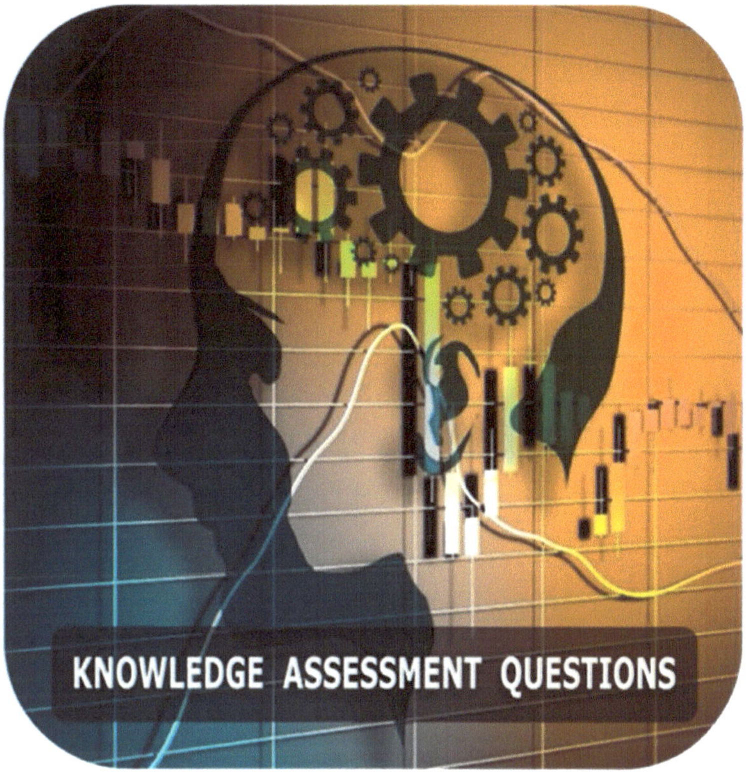

NOTICE TO STUDENTS:

The following knowledge assessment questions are presented in true / false, multiple choice, and fill-in-the-blank formats. The correct answers are provided in an Answer Key at the end of Chapter 14. These questions may or may not be presented on the quizzes and/or certification exam given by the instructor of this course.

Knowledge Assessment Questions

1) The purpose of outcome scenarios is to highlight __________ forces that may change the future, making the forces visible and easier for an analyst to recognize.

A. Peripheral

B. Secondary

C. Continuous

D. Significant

2) Analysts use four basic types of outcome scenarios: __________ scenarios, driving-force scenarios, system-change scenarios, and slice-of-time scenarios.

A. Relational

B. Hierarchy

C. Demonstration

D. Discrete

3) __________ teams are a form of challenge analysis used by intelligence teams that adopt the mindset and mentality of the adversary to become the "enemy."

A. Comparative

B. Red

C. Analytical

D. Blue

4) The technique of __________ is used to identify plausible sequences of development in a field and relate them to developments in a different field.

A. Contextual Mapping

B. Case-Based Models

C. Cross-Impact Analysis

D. Red Team Analysis

5) A __________ scenario starts at a time in the future in which a synthesis of explicit or assumed forces has already created a specific outcome.

A. Demonstration

B. Driving-Force

C. System-Change

D. Slice-of-Time

True / False

6) In an alternative future depicted by a scenario, a customer should be able to make decisions and develop strategies using outcomes developed by the analyst.

A. True

B. False

7) When analysts create a driving-force scenario, they identify the high-impact outcomes and then the plausible paths that may lead to that result.

A. True

B. False

8) For most intelligence issues, there are typically several customers interested in receiving intelligence who have different interests or questions to be answered.

A. True

B. False

9) Because scenarios are a way of understanding the dynamics driving the future, analysts attempt to identify the primary driving forces shaping the past.

A. True

B. False

10) Collection planning is a standardized and systematic methodology that derives from well-known modeling, simulation, and gaming theories used in the past.

A. True

B. False

CHAPTER 14

Considerations for Creating an Intelligence Team

Lecture Summary

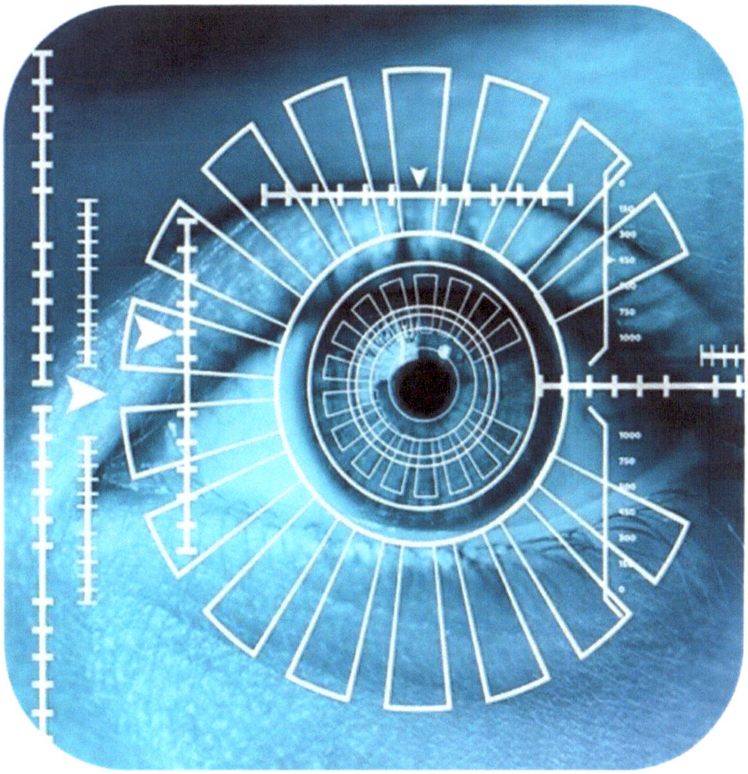

KEY KNOWLEDGE POINTS

The Threat Intelligence Analysis Team

Approaches to Building an Intelligence Team

Cognitive and Social Perspectives

Co-Acting Groups vs. Work Teams

Options for Intelligence Team Structures

Key Roles and Responsibilities

The Threat Intelligence Analysis Team

Many organizations attempt to create a *threat intelligence analysis team* without a comprehensive plan or an understanding of the needs driving the initiative. Listed below are the functions and benefits of deploying a well-planned and highly trained threat intelligence team.

1. It provides customers with a decision advantage.
2. It provides customers with a competitive advantage.
3. It influences the adversary's decision cycle.
4. It can provide understanding of complex issues.
5. It can contribute to overall operational efficiency.
6. It can contribute to enhanced security posture.
7. It supports the organization's strategic goals.

Approaches to Building an Intelligence Team

Although analysts draw on technological aids and inputs, it is ultimately the human brain that organizes and interprets data to generate assessments. A functional intelligence team involves cognitive and social processes. Analysis is a *cognitive activity* in that it is based on a system of heuristics and conclusions. Analysis is also a *social activity* in that its members draw of the expertise of their teams and support their missions. "Lone wolf" analysts – often portrayed in media – are always the exception to the rule.

The Cognitive Perspective

The different perspectives on analytic work (cognitive and social) have several significant implications for the design and leadership of intelligence teams. The *cognitive perspective* (also referred to as *co-acting groups*) puts individual analysts at the center stage. The leader organizes and encourages individual work and supports excellent individual performance. This model tends to select highly talented analysts and provides them with advanced training and technical support. In these roles, the individual analysts are held accountable for their actions.

The Social Perspective

The *social perspective* (also referred to as *interdependent work teams*) focuses on the importance of collegial interactions in competently assessing data and managing relationships with all those involved. Projects tend to be larger in size and potentially more significant. Team members have specific expertise and specialized team roles do evolve over time. In the model, the entire team produces the intelligence product and held accountable for the outcome.

Co-Acting Groups vs. Work Teams

Research and experience has identified the five conditions listed below which may increase the chances (but do not guarantee) that a Work Team will perform at peak efficiency.

Chapter 14: *Considerations for Creating an Intelligence Team*

1. *Is the group a* ***real*** *team?*
2. *Does the team have a compelling* ***purpose****?*
3. *Does the team's* ***structure*** *facilitate teamwork?*
4. *Is the team's organization* ***context*** *supportive?*
5. *Does the team have access to* ***competent*** *coaching?*

8 Shared Factors for High Performance

Whether co-acting groups or interdependent work teams, the best analytic teams are stable and bounded, with members interdependent for a particular shared purpose. Listed below are eight common factors shared by high performance teams.

1. A clear and compelling direction
2. The right size to perform the task
3. Meaningful work assignments
4. Trustworthy feedback is regularly received
5. Clear norms exist defining acceptable behavior
6. A supportive organizational context is always present
7. Members receive recognition for performance
8. Members receive coaching, training, and experience

Team Manager vs. Peer Coaching

Members of teams that are well designed experience stronger impulses to teach and assist one another than do members of co-acting groups. Managers tend to be focused on intelligence products and their attentions are focused on team outputs. They must focus on running external interference for the team, spending time dealing with resources and roadblocks. When managers step into a coaching role it is typically geared towards individual members in the form of personal and performance issues. The last priority on the list for many mangers is coaching the team *as* a team. Because of these operational realities, peer coaching remains a prominent force.

The Quality of Coaching in Teams

Spending time on coaching intelligence teams does not guarantee improvement in the individual analyst's or the team's overall performance and behavior. Coaching requires specific skillsets – not all managers are suitable coaches – and bad coaching is worse than no coaching. In this scenario, it is insufficient to demand additional coaching from the team manager. Training in coaching skills is required if it is to be effective. Managers must also ensure the coaching is well-timed and competently delivered. Despite the best training, coaching tends to be ineffectual if the team is poorly composed.

Options for Intelligence Team Structures

Experienced managers consider all options prior to settling on the one that is the most appropriate and feasible for assignment and the organizational context. Listed below are available options when choosing the most appropriate team structure.

1. **Individual Centric Teams**: These teams include *surgical teams* and *co-acting groups*.
2. **Interdependent Work Teams**: These teams can either interact face-to-face or virtually in a distributed environment.
3. **Dynamic Teams**: These team include sand dune teams assembled for specific missions and then disbanded upon completion of the mission.

Key Roles and Responsibilities

As more organizations pursue intelligence as a critical security function, many are choosing to create dedicated teams with multidisciplinary capabilities. The titles listed below represent the four primary roles in most intelligence teams.

1. *Intelligence Team Manager*
2. *Intelligence Collection Manager*
3. *Fusion Intelligence Analyst*
4. *All-Source Intelligence Analyst*

Intelligence Team Manager Responsibilities

The *intelligence team manager* accepts responsibility for all aspects of the intelligence effort and interacts with both internal and external personnel and organizations. Listed below are responsibilities typically associated with this position.

1. Develop and prepare intelligence strategies.
2. Validate the quality of team intelligence products.
3. Ensure seamless integration of all intelligence disciplines.
4. Develop, prioritize, and assign *information requirement* (IR) taskings.
5. Oversee reporting and compliance needs.
6. Mentor, train, and develop team members.
7. Manage timely and effective dissemination of intelligence products.

Intelligence Team Manager Capabilities

Intelligence team managers are required to develop innovative approaches to resolve difficult issues that can significantly impact all levels of the intelligence effort. The seven key qualities listed below are common to effective team managers.

1. The ability to push difficult actions to completion.
2. The ability to effectively resolve disagreements.
3. Possesses an understanding of intelligence and technology.
4. The ability to focus on execution of tasks.
5. Understands the nuances of internal politics.
6. Possesses a proven record of building support.
7. Possesses the skills to communicate across all disciplines.

Intelligence Collection Manager Responsibilities

Information collection management is an activity that synchronizes the planning and employment of assets in direct support of current and future operations. Listed below are responsibilities typically associated with this position.

1. Analyzes information collection requirements.
2. Evaluates available assets and capabilities.
3. Recommends taskings for collection assets.
4. Submits requests for collection support.
5. Manages timely intelligence collection and reporting.
6. Ensures all information is relevant and validated.
7. Assesses the effectiveness of all collection efforts.
8. Coordinates dissemination of information to appropriate analytical teams.

Intelligence Collection Manager Capabilities

Conducting successful information collection requires thorough and creative planning, aggressive execution, and adjustments based on the situation. The seven key qualities listed below are common to effective collection managers.

1. The ability to prioritize and focus on execution.
2. The ability to communicate with multidisciplinary teams.

3. Exceptionally attention and detail oriented.
4. Understand information inputs from multiple sources.
5. The ability to effectively summarize complex issues.
6. Possesses highly developed reasoning skills.
7. Possesses excellent written and oral presentation skills.

Fusion Intelligence Analyst Responsibilities

The *fusion intelligence analyst* synthesizes data collected from intelligence disciplines, validates data for accuracy, and converts it into a finished intelligence product. Listed below are responsibilities typically associated with this position.

1. Determines the utility of all collected data.
2. Identifies and resolves intelligence gaps.
3. Effectively communicates collector and analyst needs.
4. Refines individual collection processes as necessary.
5. Presents the final intelligence product to customers.
6. Writes *Intelligence Information Reports* (IIRs).
7. Educates intelligence teams in doctrinal methodologies.

All-Source Intelligence Analyst Responsibilities

All-source intelligence analysts analyze data from multiple sources, responds to information requests, and submits intelligence requirements in support of operations. Listed below are responsibilities typically associated with this position.

1. Clearly articulates *intelligence requirements* (IRs).
2. Formulates research questions and variables.
3. Conducts in-depth research and analysis.
4. Identifies threat decision-making processes.
5. Identifies intelligence gaps and shortfalls.
6. Accurately sources all intelligence data.
7. Confidently conveys intelligence concepts.
8. Maintains well-organized record and file systems.

Intelligence Analyst General Characteristics

Intelligence analysts can look at dissimilar objects, thoughts, or actions and see in their minds eye patterns that untrained people would otherwise overlook. Regardless of the intelligence type or implementation, the characteristics listed below are generally considered to be requirements needed for analysts to conduct successful missions.

1. Well-disciplined and highly organized
2. Communicates with clarity and credibility
3. Identify relevant patterns in chaotic noise
4. Utilize patient and methodical approaches
5. Holistically visualize at strategic levels
6. Highly flexible and adaptable to change
7. Willing to learn from previous mistakes

CHAPTER 14

Considerations for Creating an Intelligence Team

Knowledge Assessment Questions

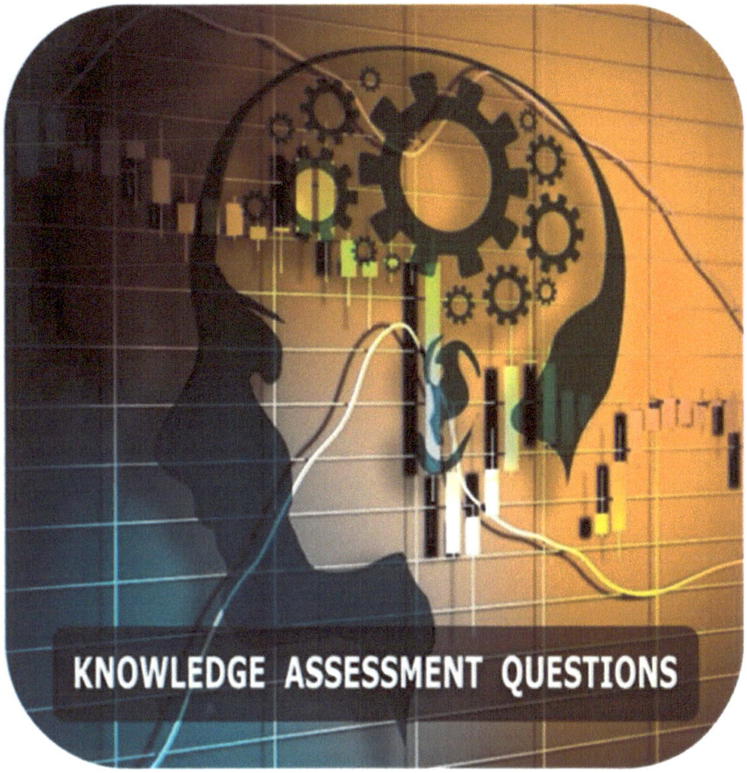

NOTICE TO STUDENTS:

The following knowledge assessment questions are presented in true / false, multiple choice, and fill-in-the-blank formats. The correct answers are provided in an Answer Key at the end of Chapter 14. These questions may or may not be presented on the quizzes and/or certification exam given by the instructor of this course.

Knowledge Assessment Questions

1) Many organizations attempt to create a threat intelligence analysis team without a __________ or an understanding of the needs driving the initiative.

A. Location

B. Survey

C. Budget

D. Plan

2) Although analysts draw on technological aids and inputs, it is ultimately the __________ that organizes and interprets data to generate assessments.

A. Technology Solution

B. Board of Directors

C. Human Brain

D. Team Leader

3) Which of the choices below is an issue that needs to be addressed when considering whether an intelligence work team will operate effectively?

A. Identified Purpose

B. Competent Coaching

C. Individual Structure

D. Individual Context

4) Which of the choices below is a characteristic of intelligence units considered to be interdependent work teams?

A. Entire Team is Held Accountable

B. Encourages Individual Work

C. Supports Individual Performance

D. Selects Highly Talented Analysts

5) Which of the choices below is a characteristic of intelligence units considered to be co-acting work groups?

A. Members are Collectively Responsible

B. Projects Tend to be Potentially Significant

C. Projects Tend to be Larger in Size

D. Individuals are Held Accountable

True / False

6) Intelligence analysts can look at similar objects, thoughts, or actions and see in their minds eye patterns that untrained people would otherwise notice.

A. True

B. False

7) The different perspectives on analytic work (cognitive and social) have several significant implications for the design and leadership of intelligence teams.

A. True

B. False

8) The all-source analyst synthesizes data collected from intelligence disciplines, validates data for accuracy, and converts it into a finished intelligence product.

A. True

B. False

9) The collection manager accepts responsibility for all aspects of the intelligence effort and interacts with both internal and external personnel and organizations.

A. True

B. False

10) Experienced managers consider all options prior to settling on the one that is the most appropriate and feasible for assignment and the organizational context.

A. True

B. False

Knowledge Assessment Questions
Answer Key

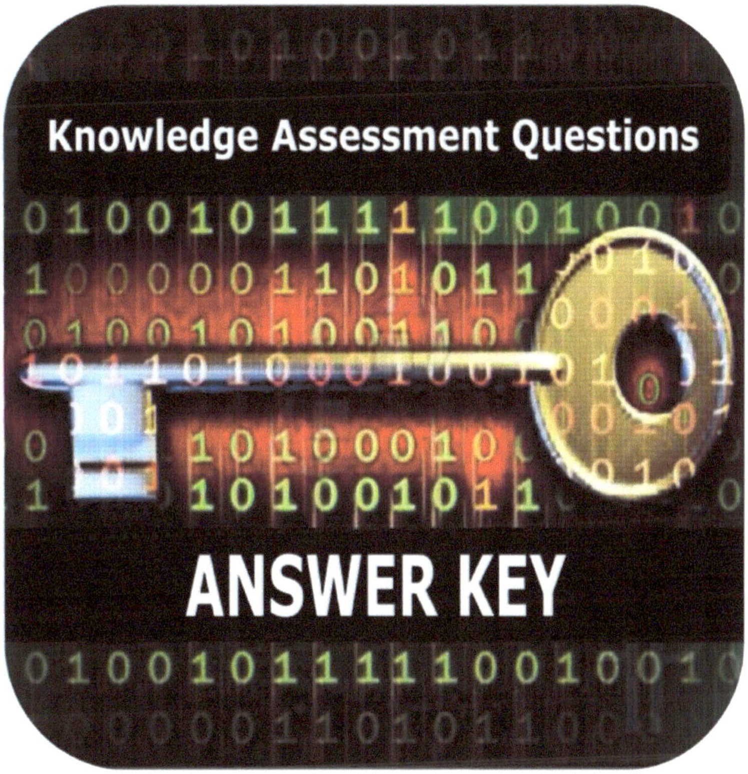

NOTICE TO STUDENTS:

The following answers to the domain knowledge assessment questions are presented as true / false, multiple choice, and fill-in-the-blank. The correct answers to the true / false questions are notated in blue. The questions presented throughout this course may or may not be included on the actual certification exam.

Knowledge Assessment Answer Key

1) Describing intelligence by type can be defined by implementation levels such as __________, operational, and strategic.

A. Tactical

B. Mission

C. Technical

D. Asymmetric

2) The *OODA Loop*, developed by Colonel John Boyd, is a recognitional decision-making model that includes the actions observe, orient, decide, and __________ as part of the cycle.

A. Assess

B. Recognize

C. Analyze

D. Act

3) What choice below matches the following definition: "*Intelligence-gathering by interception of signals, whether communications between people or from electronic signals not directly used in communication*."

A. HUMINT

B. COMINT

C. SIGINT

D. GEOINT

4) A SWOT analysis is a technique used to help an organization identify __________, weaknesses, opportunities, and threats related to strategic intelligence and operational planning.

A. Situations

B. Similarities

C. Strengths

D. Symptoms

5) Which of the choices listed below would be a goal of an analyst engaged in an operational intelligence action against a threat activity?

A. Interrogating

B. Targeting

C. Dissuading

D. Directing

True / False

6) Modern intelligence activities involve the planning and coordination of multiple organizational units and are executed by teams, not individuals.

A. True

7) Most of the information collected by agencies and organizations in intelligence operations is classified, and usually available to all who seek it out.

B. False

Most of the information collected by agencies and organizations in intelligence operations is open-source, and usually available to all who seek it out.

8) Although intelligence provides a great advantage in supporting organizational interests, there are significant limitations inherent to it as well.

A. True

9) Strategic intelligence refers to quick-reaction intelligence that supports ongoing operations by identifying immediate opportunities and threats.

B. False

Tactical intelligence refers to quick-reaction intelligence that supports ongoing operations by identifying immediate opportunities and threats.

10) Intelligence operations are often considered asymmetric activities, allowing weaker organizations to compensate for their shortcomings and vulnerabilities.

A. True

Knowledge Assessment Answer Key

1) The __________ phase of the intelligence cycle is comprised of all activities that identify information requirements and ensure requirements are met.

A. Analysis and Production

B. Planning and Direction

C. Utilization and Feedback

D. Intelligence Processing

2) In the __________ intelligence life cycle process, collectors and customers can not only see into the process but have many more opportunities to contribute to it.

A. Traditional

B. Customer-Centric

C. Target-Centric

D. Community-Centric

3) In the network-centric intelligence model, systems are viewed as individual targets and networks are comprised of __________ and links.

A. Connections

B. Cells

C. Edges

D. Nodes

4) Object-based production in the target-centric intelligence model organizes the analyst's efforts around objects (targets) and features a __________ sharing of knowledge.

A. Cloud-Based

B. Clearance-Based

C. Security-Based

D. Agency-Based

5) The __________ phase acquires and provisions information to be provided to teams in the next chain of the traditional intelligence cycle.

A. Intelligence Processing

B. Utilization and Feedback

C. Planning and Direction

D. Intelligence Collection

True / False

6) Although most of the focus is on collection and operations, most of the failures in intelligence are due to inadequate or nonexistent funding.

B. False

Although most of the focus is on collection and operations, most of the failures in intelligence are due to inadequate or nonexistent analysis.

7) The goal of intelligence is to develop knowledge of the adversary network, make the analyst's network effective, and to render the adversary network ineffective.

A. True

8) In addition to being a system and a network, targets of intelligence interest have special attributes and technical attributes that must be considered by analysts.

B. False

In addition to being a system and a network, targets of intelligence interest have spatial attributes and temporal attributes that must be considered by analysts.

9) It may be easier to see the adversary's side as a network than to see that one's own intelligence assets form a network and to fully exploit its strengths.

A. True

10) In the traditional intelligence process, collectors and customers can not only see into the process but have many more opportunities to contribute to it.

B. False

In the target-centric intelligence process, collectors and customers can not only see into the process but have many more opportunities to contribute to it.

Knowledge Assessment Answer Key

1) What choice below is a result of passing the *Electronic Communications Privacy Act* (ECPA) in 1986, legislation intended to expand and revise wiretapping and electronic eavesdropping provisions?

A. Created the *Wiretap Act*

B. Created the *Stored Communications Act*

C. Created the *Pen Register Act*

D. Expanded Electronic Eavesdropping

2) What type of hacker matches the following definition: "*A person who is not technologically sophisticated, who randomly seeks out a specific weakness to gain access to a network or system*"?

A. White Hat

B. Script Kiddie

C. Gray Hat

D. Ethical

3) What hacktivist group matches the following definition: "*A decentralized and international hacktivist group known for its cyber-attacks against governments, institutions, agencies, and corporations*"?

A. Anonymous

B. Elfin Team

C. Ocean Lotus

D. Fancy Bear

4) Ideas become __________ when they are physical and tangible works that may be accessed and viewed by others.

A. Profitable

B. Commercially Available

C. Time Sensitive

D. Intellectual Property

5) A(n) __________ is a threat actor, typically a nation-state or state-sponsored group, attempting to gain unauthorized access to a network.

A. Hacktivist Group

B. Script Kiddie

C. Advanced Persistent Threat

D. Nonstate Hacker

True / False

6) The ability to communicate with others of diverse backgrounds ensures that individuals can be steadily introduced to a movement's principles.

A. True

7) An abstraction involves the modification of technology in order to allow it to be used in unintended ways for both legal and illegal purposes.

B. False

A hack involves the modification of technology in order to allow it to be used in unintended ways for both legal and illegal purposes

8) Threat intelligence can be derived from external sources such as open-source platforms, or communications between threat information sharing groups.

A. True

9) The Fourth Amendment protects people from self-incrimination, meaning giving a statement that might implicate oneself in a crime.

B. False

The Fifth Amendment protects people from self-incrimination, meaning giving a statement that might implicate oneself in a crime

10) Evolving privacy and data protection regulations across geographies may add to the streamlining of gathering actionable intelligence and accessing sources.

B. False

Evolving privacy and data protection regulations across geographies may add to the complexity of gathering actionable intelligence and accessing sources

Knowledge Assessment Answer Key

1) What choice below is one of the three basic thinking abilities required by intelligence analysts to be successful in the performance of their duties?

A. Logical Recognition

B. Dependency on Memory

C. Premature Closure

D. Information Ordering

2) __________ reasoning is a process of discovery in which an analyst establishes a relationship between events under observation or study.

A. Inductive

B. Abductive

C. Deductive

D. Analogical

3) A(n) __________ is a disproportionate weight in favor of or against an idea or thing, usually in a way that is closed-minded, prejudicial, or unsupported by fact.

A. Intellectual Standard

B. Bias

C. Element of Thought

D. Logical Fallacy

4) __________ reasoning is a logical process in which a conclusion is based on the concordance of multiple premises that are generally assumed to be true.

A. Analogical

B. Inductive

C. Deductive

D. Abductive

5) When analysts dissect their thinking and examine its parts, they use standards of quality referred to as __________ or *standards of thought*.

A. Elements of Thought

B. Critical Thinking

C. Logical Reasoning

D. Intellectual Standards

True / False

6) *Fallacies of omission* occur when an analyst leaves out necessary material (either intentionally or unintentionally) in a conclusion or inference.

A. True

7) While the *elements of thought* provide a framework for qualifying thinking, the standards of thought provide criteria to assess the analysis of thinking.

B. False

While the elements of thought provide a framework for analyzing thinking, the standards of thought provide criteria to assess the quality of thinking

8) When communicating the results of analysis, the analyst must ensure that the collection manager understands the conclusions and how they were reached.

B. False

When communicating the results of analysis, the analyst must ensure that the customer understands the conclusions and how they were reached

9) Analysts detect and impose patterns on apparently random entities and events in order to understand them, often doing this without being aware of it.

A. True

10) Abductive reasoning is a method of processing information that compares the similarities between new concepts and previously understood concepts.

B. False

Analogical reasoning is a method of processing information that compares the similarities between new concepts and previously understood concepts

Knowledge Assessment Answer Key

1) In the field of intelligence, the facilitator between the producers of intelligence (analysts) and the operators (collectors) is the __________.

A. Collection Manager

B. Fusion Analyst

C. Team Leader

D. Intelligence Customer

2) The benefit of communicating "__________ intent" is the holistic perspective it provides internal and external collectors isolated from the planning process.

A. Processing

B. Collection

C. Exploitation

D. Dissemination

3) When true collection __________ have been defined, it is time to plan for collection to fill them and offer guidance to the sources of collection.

A. Plans

B. Methods

C. Gaps

D. Sources

4) Collection managers cannot rely solely on previous strategies because predictable collection methods can be __________.

A. Repeated

B. Modelled

C. Implemented

D. Defeated

5) When collection managers are developing the initial collection plan, it is imperative that customers __________ their specific needs.

A. Articulate

B. Question

C. Refine

D. Support

True / False

6) The Collection Manager operates across all echelons, all security and network domains, and across all computing environments to aid intelligence personnel.

A. True

7) Identified sources need to be made explicit, and the intelligence collection effort must fill in the missing pieces of the model to answer the customer's questions.

B. False

Identified gaps need to be made explicit, and the intelligence collection effort must fill in the missing pieces of the model to answer the customer's questions.

8) Intelligence collection requirements fall within two categories: tasking external resources belonging to the organization or requesting use of future resources.

B. False

Intelligence collection requirements fall within two categories: tasking organic resources belonging to the organization or requesting use of external resources.

9) Most collection strategies are developed as a result of past experiences; use of existing methods saves time, effort, and organizational resources.

A. True

10) Whenever possible, write into the *Specific Information Requirements* the requirement for direct dissemination of intelligence to the customer of the product.

B. False

Whenever possible, write into the *Specific Orders and Requests* the requirement for direct dissemination of intelligence to the customer of the product.

Knowledge Assessment Answer Key

1) On large and complex intelligence projects, the __________ statement will itself become a formal product to be disseminated to stakeholders.

A. Qualitative

B. Parameter

C. Quantitative

D. Implementation

2) When analysts receive the initial tasking (from a customer of self-initiated) time pressures may tempt them to give the __________ definition minimal consideration.

A. Reporting

B. Collection

C. Issue

D. Reasoning

3) What choice listed below is a factor to be considered by an intelligence analyst when defining the requirements for the *Systems Acquisition and Implementation* issue category?

A. Key Actors

B. Implementations

C. Utilities

D. Readiness

4) A __________ is a classification system in which objects are arranged into natural or related groups based on a factor common to each object in the group.

A. Taxonomy

B. Doctrine

C. Ontology

D. Ideology

5) A __________ is a hypothesis that has been accepted as true and will be a part of the problem definition or the final intelligence assessment product.

A. Logical Fallacy

B. Key Assumption

C. Issue Definition

D. Cultural Bias

True / False

6) All intelligence analysis efforts begin with an issue or problem definition which aligns with the customer's interest about the target.

A. True

7) Target deconstruction is the technique for taking a seemingly intractable problem and breaking is down into a series of manageable sub-problems.

B. False

Issue decomposition is the technique for taking a seemingly intractable problem and breaking is down into a series of manageable sub-problems

8) Having insight into an analyst's priorities makes it easier to establish valid assumptions regarding their requirements for intelligence issues.

B. False

Having insight into a customer's priorities makes it easier to establish valid assumptions regarding their requirements for intelligence issues.

9) A well-defined statement of the problem (the result of these answers) still requires two additional steps: issue definition and issue decomposition.

A. True

10) The PMESII categories are environmental factors used to analyze the cultural and social conditions in which the operational variables (ASCOPE) exist.

B. False

The ASCOPE categories are environmental factors used to analyze the cultural and social conditions in which the operational variables (PMESII) exist.

Knowledge Assessment Answer Key

1) When working with intelligence modeling concepts, analysts typically use __________ models to prescribe a course of action for an issue.

A. Normative

B. Conceptual

C. Descriptive

D. Physical

2) When working with descriptive models, analysts use __________ models when the issue or problem contains uncertainties or probabilities.

A. Deterministic

B. Stochastic

C. Linear

D. Nonlinear

3) It is an established standard in intelligence that, if the organization can afford the resources, independent groups should provide __________ analysis.

A. Supporting

B. Accurate

C. Competing

D. Speculative

4) Analysts creating target frameworks for analytical work use __________ system models when describing actors, objects, or relationships.

A. Elemental

B. Process

C. Functional

D. Structural

5) The purpose of creating intelligence __________ is to help predict what the profiled person or group will most likely do in various circumstances and conditions.

A. Profiles

B. Cases

C. Models

D. Frameworks

True / False

6) Modeling does not always require a computer, and many useful models that analysts can utilize to solve complex issues exist only on paper.

A. True

7) Coordinating techniques are a simple but useful form of modeling available to analysts that do not require sophisticated simulation exercises.

B. False

Comparative techniques are a simple but useful form of modeling available to analysts that do not require sophisticated simulation exercises.

8) Competitive modeling is a critical element of all intelligence as leaders, terrorists, and criminals have a modus operandi (a standard operational pattern).

B. False

Pattern modeling is a critical element of all intelligence as leaders, terrorists, and criminals have a modus operandi (a standard operational pattern).

9) The generic target framework is the starting point for any analysis, as defining and deconstructing the target framework is the most important step in analysis.

A. True

10) Starting from the low-level view of the target, the analyst will deconstruct it into mid-categories (elements) and then higher-elements (all represented by models).

B. False

Starting from the top-level view of the target, the analyst will deconstruct it into sub-categories (elements) and then sub-elements (all represented by models).

Knowledge Assessment Answer Key

1) In intelligence analysis, __________ is the organizing of relevant information in a coherent way, taking both the source and the context into consideration.

A. Collection

B. Collation

C. Comparison

D. Corroboration

2) __________ analysis uses new data to modify previously estimated probabilities and can therefore be used to narrow the error bounds on estimates.

A. Theoretical

B. Historical

C. Bayesian

D. Substantive

3) Intelligence analysts working with __________ evidence are considering two data sets that favor different conclusions.

A. Corroborative

B. Convergent

C. Cumulative

D. Divergent

4) The fundamental objective in weighing intelligence evidence is determining its __________; the analyst must evaluate it for completeness and veracity.

A. Credibility

B. Source

C. Channel

D. Method

5) It is often important to combine evidence and demonstrate the logical process of reaching a conclusion based on that evidence through __________.

A. Literal Intelligence

B. Structured Argumentation

C. Professional Sources

D. Occam's Razor

True / False

6) A significant consequence of the information explosion is that analysts must now deal with "big data" when collating and analyzing intelligence products.

A. True

7) The intelligence analysis process involves combining identical types of evidence and helps analysts to make qualitative judgments about conclusions to support.

B. False

The intelligence analysis process involves combining disparate types of evidence and helps analysts to make qualitative judgments about conclusions to support.

8) Prior to starting an intelligence collection effort, analysts should ensure that they are already aware of information already collected on a hypothesis.

B. False

Prior to starting an intelligence collection effort, analysts should ensure that they are already aware of information already collected on a target.

9) The issue decomposition and the target framework are separate concepts and separate models, but they are closely related to each other.

A. True

10) The information reliability is rated between "A" (complete reliability) to "E" (invalid information), with "F" for a source without a sufficient history to rate accurately.

B. False

The source reliability is rated between "A" (complete reliability) to "E" (invalid information), with "F" for a source without a sufficient history to rate accurately.

Knowledge Assessment Answer Key

1) Analysts conducting anticipatory analysis must consider __________, a background force that addresses the tendency to resist change.

A. Opposition

B. Inertia

C. Contamination

D. Feedback

2) __________ is an estimative approach used by analysts in anticipatory analysis that assumes forces affecting the target do change.

A. Extrapolation

B. Forecasting

C. Projection

D. Prediction

3) __________ diagrams represent a systematic approach to force analysis and present a unique form of causal model for the intelligence analyst and customer.

A. Premature Closure

B. Influece Net

C. Demonstration Scenario

D. Influence Tree

4) Analysts conducting anticipatory analysis must consider __________, a background force that addresses the degradation of any external factor.

A. Contamination

B. Prediction

C. Opposition

D. Inertia

5) A primary objective of __________ in intelligence operations is to define possible alternative outcomes of the target framework, not just the most likely ones.

A. Projection

B. Forecasting

C. Extrapolation

D. Prediction

True / False

6) Estimating the future state of a target can only be done using probabilities, so the risk of a mistake is greater than when analysts rely on established facts.

A. True

7) Target models depict elements of the target, elements affecting the target, and describe how those elements are related by probabilities.

B. False

Causal models depict elements of the target, elements affecting the target, and describe how those elements are related by probabilities.

8) Intelligence analysts should advise customers that forecasts are static and need constant adjustment to be helpful in guiding thoughts and actions.

B. False

Intelligence analysts should advise customers that forecasts are dynamic and need constant adjustment to be helpful in guiding thoughts and actions.

9) *Influence net* modeling is a powerful tool for projection of complex target models where the influence tree would be too cumbersome for practical analytical use.

A. True

10) The first step in probabilistic reasoning is formally stating that more than one outcome is possible (alternatives) by listing all foreseeable possible scenarios.

A. True

Knowledge Assessment Answer Key

1) __________ is an effective structured analytic technique when information elements can be broken out into categories for comparison using automated computer software.

A. Event Mapping

B. Event Trees

C. Matrices

D. Sorting

2) __________ probability does not determine a timeline for events; it attempts to allow analysts to predict an outcome by applying a percentage of probability.

A. Event Map

B. Event Tree

C. Subjective

D. Quantitative

3) An __________ is a graphic depiction of a possible sequence of events, including junctures within the event sequence, depicting threat options and decisions.

A. Event Map

B. Event Tree

C. Matrix

D. Qualitative Dataset

4) __________ are useful whenever there are more options or more intricate data than can be conceptualized at one time without a visual representation.

A. Matrices

B. Demonstration Scenario

C. Event Map

D. Event Tree

5) __________ uses a brainstorming diagram to represent the scenarios in the hypotheses linked around a central phrase representing the issue to be analyzed.

A. Extrapolation

B. Subjective Probability

C. Forecasting

D. Event Mapping

True / False

6) The three-dimensional design of matrices increases their use for collating data on complex issues and leaving out pertinent data will oversimplify issues.

B. False

The two-dimensional design of matrices limits their use for collating data on complex issues and leaving out pertinent data will oversimplify issues.

7) Sorting massive amounts of data can provide insights into trends or anomalies that warrant further investigation that may have otherwise gone unnoticed.

A. True

8) The intelligence team matrix is an analytic technique used by intelligence analysts to efficiently look at information from the threat's point of view.

B. False

The threat intentions matrix is an analytic technique used by intelligence analysts to efficiently look at information from the threat's point of view.

9) It is important to enter the information for every decision-making criterion in a column of a threat intentions matrix before moving to the next to mitigate hypotheses.

B. False

It is important to enter the information for every decision-making criterion in a column of a threat intentions matrix before moving to the next to mitigate bias.

10) Subjective probability is a quantitative expression of an analyst's degree of belief in the truth of a statement relative to all other alternative possibilities.

A. True

Knowledge Assessment Answer Key

1) There are four levels of relationship models, each using increasing sophisticated analytic approaches: hierarchy, link, __________, and network models.

A. Host

B. Node

C. Cell

D. Matrix

2) __________ modeling is routinely used in cybersecurity and law enforcement to identify Advanced Persistent Threats, Dark Web criminals, and terrorist organizations.

A. Matrix

B. Node

C. Link

D. Cell

3) Target network analysis has become one of the principal tools for dealing with complex systems as a result of __________-based analytic methods and software.

A. Relationship

B. Computer

C. Analysis

D. Threat

4) __________ in a target network can include people, places, objects, and organizations, all of which can be treated as separate networks if required.

A. Nodes

B. Links

C. Edges

D. Domains

5) __________ Analysis is a method of identifying and prioritizing High-Value Targets (HVTs) and High-Payoff Targets (HPTs) to determine which should be engaged.

A. Matrix

B. Computational

C. Network

D. Target Value

True / False

6) The network model of multidimensional conflict between opposing networks is increasingly applicable to all intelligence and the network analysis tool of choice.

B. False

The *netwar* model of multidimensional conflict between opposing networks is increasingly applicable to all intelligence and the network analysis tool of choice.

7) There are numerous types of networks but for practical purposes, they are all target networks, even though the network is not necessarily hostile.

A. True

8) A premise of threat network modeling is that these networks have strengths that can be exploited once they are identified by intelligence analysts.

B. False

A premise of threat network modeling is that these networks have vulnerabilities that can be exploited once they are identified by intelligence analysts.

9) Engaging targets that the adversary can least afford to lose deprives them of their functionality and forces them to alter their established funding cycle.

B. False

Engaging targets that the adversary can least afford to lose deprives them of their functionality and forces them to alter their established decision cycle.

10) Equivalence refers to the sources and distribution of power in a social structure, and the network perspective suggests power is derived from relationships.

B. False

Centrality refers to the sources and distribution of power in a social structure, and the network perspective suggests power is derived from relationships.

Knowledge Assessment Answer Key

1) The challenge of a __________ simulation is to select the right uncertainties for the input variables which represent the correct probability distribution.

A. Stochastic

B. Deterministic

C. Continuous

D. Hybrid

2) Intelligence analysts use __________ simulation models extensively to assess the performance of adversarial threats, capabilities, and even financial economies.

A. Snapshot

B. Interactive

C. Static

D. Intelligence

3) __________ analyze the decision process of two or more parties in conflict or cooperation and is valuable in simulation for its use of mathematical equations.

A. Collective Decision-Making Models

B. Emotional Models

C. Game Theory Models

D. Cultural Models

4) The tendency of leaders to adopt suboptimal choices leads to a variant of the rational decision-making simulation referred to as the __________ model.

A. Cultural

B. Emotional

C. Rational

D. Administrative

5) __________ models are used to simulate the performance of an entire economy or any segment of it such as competitive industries.

A. Economic

B. Geospatial

C. Political

D. Social

True / False

6) The simulation models typically used by intelligence analysts are classified as continuous simulations, discrete-event simulations, or a hybrid of the two.

A. True

7) Composite simulations are a powerful tool used by intelligence analysts for modeling and analyzing threat, neutral, and allied networks.

B. False

Network simulations are a powerful tool used by intelligence analysts for modeling and analyzing threat, neutral, and allied networks.

8) An important aspect of the decision process to consider when analyzing behavior is the administrative one, a dominant component in many influential leaders.

B. False

An important aspect of the decision process to consider when analyzing behavior is the emotional one, a dominant component in many influential leaders.

9) Rational decision-making requires the systematic evaluation of costs or benefits accruing to courses of action that may be taken by an individual.

A. True

10) One of the most powerful applications of geospatial simulations comes from combining them with models such as *Operational Requirements Analysis* (ORA).

A. True

Knowledge Assessment Answer Key

1) The purpose of outcome scenarios is to highlight __________ forces that may change the future, making the forces visible and easier for an analyst to recognize.

A. Peripheral

B. Secondary

C. Continuous

D. Significant

2) Analysts use four basic types of outcome scenarios: __________ scenarios, driving-force scenarios, system-change scenarios, and slice-of-time scenarios.

A. Relational

B. Hierarchy

C. Demonstration

D. Discrete

3) __________ teams are a form of challenge analysis used by intelligence teams that adopt the mindset and mentality of the adversary to become the "enemy."

A. Comparative

B. Red

C. Analytical

D. Blue

4) The technique of __________ is used to identify plausible sequences of development in a field and relate them to developments in a different field.

A. Contextual Mapping

B. Case-Based Models

C. Cross-Impact Analysis

D. Red Team Analysis

5) A __________ scenario starts at a time in the future in which a synthesis of explicit or assumed forces has already created a specific outcome.

A. Demonstration

B. Driving-Force

C. System-Change

D. Slice-of-Time

True / False

6) In an alternative future depicted by a scenario, a customer should be able to make decisions and develop strategies using outcomes developed by the analyst.

A. True

7) When analysts create a driving-force scenario, they identify the high-impact outcomes and then the plausible paths that may lead to that result.

B. False

When analysts create a demonstration scenario, they identify the high-impact outcomes and then the plausible paths that may lead to that result.

8) For most intelligence issues, there are typically several customers interested in receiving intelligence who have different interests or questions to be answered.

A. True

9) Because scenarios are a way of understanding the dynamics driving the future, analysts attempt to identify the primary driving forces shaping the past.

B. False

Because scenarios are a way of understanding the dynamics driving the future, analysts attempt to identify the primary driving forces shaping the present.

10) Collection planning is a standardized and systematic methodology that derives from well-known modeling, simulation, and gaming theories used in the past.

B. False

Scenario planning is a standardized and systematic methodology that derives from well-known modeling, simulation, and gaming theories used in the past.

Knowledge Assessment Answer Key

1) Many organizations attempt to create a threat intelligence analysis team without a __________ or an understanding of the needs driving the initiative.

A. Location

B. Survey

C. Budget

D. Plan

2) Although analysts draw on technological aids and inputs, it is ultimately the __________ that organizes and interprets data to generate assessments.

A. Technology Solution

B. Board of Directors

C. Human Brain

D. Team Leader

3) Which of the choices below is an issue that needs to be addressed when considering whether an intelligence work team will operate effectively?

A. Identified Purpose

B. Competent Coaching

C. Individual Structure

D. Individual Context

4) Which of the choices below is a characteristic of intelligence units considered to be interdependent work teams?

A. Entire Team is Held Accountable

B. Encourages Individual Work

C. Supports Individual Performance

D. Selects Highly Talented Analysts

5) Which of the choices below is a characteristic of intelligence units considered to be co-acting work groups?

A. Members are Collectively Responsible

B. Projects Tend to be Potentially Significant

C. Projects Tend to be Larger in Size

D. Individuals are Held Accountable

True / False

6) Intelligence analysts can look at similar objects, thoughts, or actions and see in their minds eye patterns that untrained people would otherwise notice.

B. False

Intelligence analysts can look at dissimilar objects, thoughts, or actions and see in their minds eye patterns that untrained people would otherwise overlook.

7) The different perspectives on analytic work (cognitive and social) have several significant implications for the design and leadership of intelligence teams.

A. True

8) The all-source analyst synthesizes data collected from intelligence disciplines, validates data for accuracy, and converts it into a finished intelligence product.

B. False

The fusion analyst synthesizes data collected from intelligence disciplines, validates data for accuracy, and converts it into a finished intelligence product.

9) The collection manager accepts responsibility for all aspects of the intelligence effort and interacts with both internal and external personnel and organizations.

B. False

The team manager accepts responsibility for all aspects of the intelligence effort and interacts with both internal and external personnel and organizations.

10) Experienced managers consider all options prior to settling on the one that is the most appropriate and feasible for assignment and the organizational context.

A. True

Course Lecture Slides and Notes

NOTICE TO STUDENTS:

This section of the course workbook contains all the lecture slides presented in this course, and designated space next to each slide to make notes during the presentation. You are encouraged to take detailed notes to augment the information on the slides and use these notes as memory aids during the certification exam.

INTRODUCTION TO

Intelligence Analysis and Cyber Threat Modeling

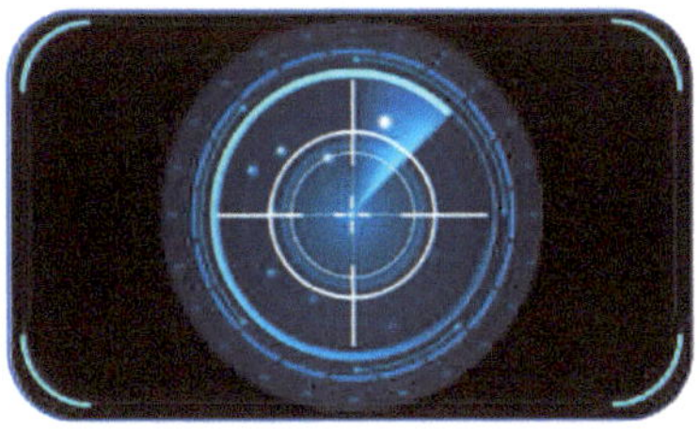

Your Cybersecurity Career Starts Here®

PHASE2 ADVANTAGE © 2023 +1.912.335.2217

This cybersecurity training program is being presented by

PHASE2 ADVANTAGE

Phase2Advantage.com

PHASE2 ADVANTAGE © 2023 +1.912.335.2217

About the Course Designer

Michael I. Kaplan

- Founder and CEO, Phase2 Advantage
- Chair, STC Cybersecurity Advisory Board
- US Army: Intelligence and Special Operations
- Federal Fugitive Task Force (International)
- linkedin.com/in/michaelikaplan
- michael.kaplan@phase2advantage.com

PHASE2 ADVANTAGE © 2023 +1.912.335.2217

The Purpose and Value of Intelligence

PHASE2 ADVANTAGE © 2023 CHAPTER 01

Intelligence Operations

Definition:

The variety of intelligence and counterintelligence tasks that are carried out by various organizations, and activities within the intelligence process including planning and direction; collection; analysis and production; processing and exploitation; dissemination; and evaluation and feedback.

Dictionary of Military and Associated Terms
jcs.mil/Portals/36/Documents/Doctrine/pubs/dictionary.pdf

PHASE2 ADVANTAGE © 2023 CHAPTER 01

The Intelligence Analyst

Definition:

An individual that reviews information from multiple sources, disciplines, and perspectives in support of intelligence planning and operations; evaluates and synthesizes large quantities of data (which may be fragmented and contradictory) into high quality, fused targeting and intelligence products; recognizes and mitigates cognitive biases and factual errors which may affect the analysis process; and recommends analytic approaches or solutions to problems and situations for which information is incomplete or for which no precedent exists.

NICE Cybersecurity Workforce Framework
niccs.cisa.gov/workforce-development/cyber-security-workforce-framework

PHASE2 ADVANTAGE © 2023 CHAPTER 01

Modern Intelligence: The Popular Myth

Modern intelligence activities involve the planning and coordination of multiple organizational units and are executed by *teams*, **not** *individuals*.

PHASE2 ADVANTAGE © 2023 CHAPTER 01

Describing Intelligence by Type (I)

- **HUMINT**: Human Intelligence
- **COMINT**: Communications Intelligence
- **GEOINT**: Geospatial Intelligence
- **SIGINT**: Signals Intelligence
- **OSINT**: Open-Source Intelligence

Intelligence is often described in terms of the methods used in its **collection**, the gathering of various types of data through a variety of sources.

PHASE2 ADVANTAGE © 2023 CHAPTER 01

Describing Intelligence by Type (II)

- Defined in Terms of the **Analytic Process**
- - Conversion Process to a Finished Product
- Defined in Terms of **Implementation Level**
- - *Tactical*, *Operational*, and *Strategic*
- Defined in Terms of the **Mission Type**
- - *Indication & Warning*, *Prediction*, *Deception*
- Defined by the **Type of Category**

Most of the information collected by agencies and organizations in intelligence operations is open-source, and usually available to all who seek it out.

PHASE2 ADVANTAGE © 2023 CHAPTER 01

The Primary Mission: Decision Advantage

- **Decision Advantage**
- - This can Prove to be a Challenging Task
- **1**: Access an Adversary's **Decision Cycle**
- **2**: Alter the Decision-Making Process
- **3**: Shift the Adversary's Preferred Outcomes
- **4**: Alter the Adversary's Target Models

Decision advantage is the ability to find out an adversary's intentions, and then enabling action to preempt, prevent, frustrate, or disrupt those plans.

PHASE2 ADVANTAGE © 2023 CHAPTER 01

The OODA Loop: **Decision Cycle**

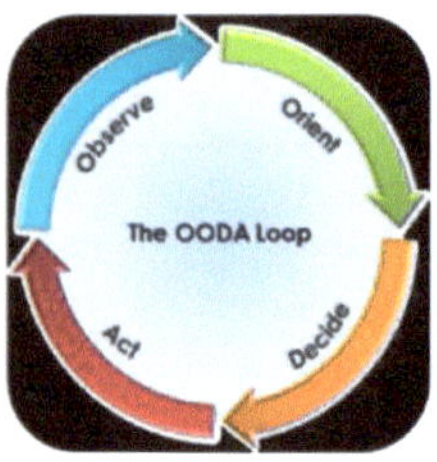

- Developed by Col. John Boyd (USAF)
- Recognitional Decision-Making Model
- **Observe**
- **Orient**
- **Decide**
- **Act**
- ***Evaluate*** (Added in Newer Models)

The organization or adversary who has the most *decision advantage* is the one who can cycle through the *OODA Loop* the fastest and do it correctly.

PHASE2 ADVANTAGE © 2023 CHAPTER 01

Analysis Scenario (ROE)

ROE 1:

No actions may be taken to alter the environment or any of the components presented in this scenario. Nothing may be added, removed, or manipulated by participants in this analysis. All facts needed have been presented; there is nothing "hidden."

ROE 2:

Participants will have 5 minutes to read the scenario, conduct their analysis, and be prepared to submit their answer based on the Information Requirement (IR) Tasking Order.

ROE 3:

If you believe you have solved the IR Tasking Order, write it down and stand by. Do not discuss the answer until the end of the exercise.

ROE:
Rules of Engagement

PHASE2 ADVANTAGE © 2023 CHAPTER 01

Analysis Scenario

Situation:

Two men are standing in the middle of a room facing in opposite directions; one facing East, the other facing West. The man facing East is 6'4", the man facing West is 5'08." The men are not allowed to move at all.

The room is 10'x10'x10' and well-lit by a lighting fixture in the ceiling. The walls and ceiling are white. There is a 12"x12" mirror on the South wall, and a 12"x12" window on the North wall.

The 2 men in the room cannot see the mirror or window in their peripheral vision, nor can they see any reflections cast by the mirror, window, or lighting fixture.

IR Tasking Order:

Determine what actions need to be taken that will allow the two men described in the scenario to be able to see each other.

Final Analysis: Intelligence Product

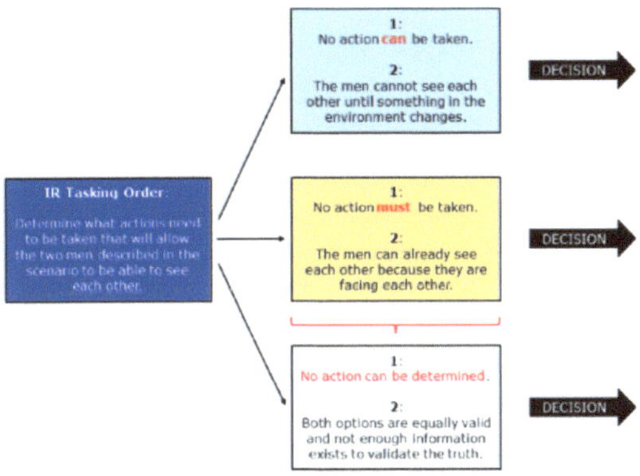

Common Analytical Errors

- Improper **framing** of the situational context.
- Improper **issue decomposition** to identify sub-issues.
- Failure to identify **alternative competing hypotheses** (ACHs).
- Failure to apply the principle of **Occam's Razor**.
- Facilitating **confirmation bias** with irrelevant data.
- Limiting the analysis with **premature closure** due to lack of options.
- Failure to separate facts from **opinions and assumptions** during the verbal portion of the IR Tasking Order.

Typical goals of intelligence are to establish facts and develop precise, reliable, and valid inferences for use in strategic decision making and operational planning.

Intelligence Sources and Methods

- People and Tools used to Collect Information
- People and Tools used to Conduct Operations
- As Important as the Information Provided
- Improved Outcomes are not Guaranteed
- Activities must Further Strategic Interests
- Effective Intelligence is a Force Multiplier

Intelligence operations are often considered **asymmetric** activities, allowing weaker organizations to compensate for their shortcomings and vulnerabilities.

Strategic Intelligence

- Deals with Long-Range Issues
- Produced for Senior Leadership
- - Strategic Planning and Budgets
- - Technology Development Plans
- - Overall Security Posture of the Organization
- **SWOT Analysis**
- Strength, Weakness, Opportunities, Threats

Strategic intelligence is best understood in terms of the methodology used in strategic planning known as "SWOT."

Operational Intelligence (I)

- Primary Concerns of Operational Intelligence
- - **Detecting** Threat Activity
- - **Investigating** Threat Activity
- - **Identifying** Threat Activity
- - **Targeting** Threat Activity
- - **Disrupting** Threat Activity
- - **Defeating** Threat Activity

Operational intelligence focuses on the capabilities and intentions of both known adversaries and unknown potential adversaries.

Operational Intelligence (II)

- **Predictive Intelligence is Preferred**
- - Model the Adversary's Tools and Tactics
- - Assess the Impacts of Topology and Assets
- - Changes to Tactics Based on Environment
- **Predicting Future Actions is Difficult**
- - Complete Information is Always Lacking
- - Adversary may use Denial and Deception

Operational intelligence is used by national-level and international business organizations to support and plan for preemptive cybersecurity programs.

Tactical Intelligence

- Used at the Front Line of any Conflict
- Provides Functional Damage Assessments
- - Disruption of the Target's Effectiveness
- Contributes to Successful Investigations
- Driven by the Need for Fast Response
- Referred to as **Current Intelligence**
- Used Every Day in Many Situations

Tactical intelligence refers to quick-reaction intelligence that supports ongoing operations by identifying immediate opportunities and threats (SW**OT**).

The Limitations of Intelligence

- **Some Things are Simply Unknowable**
- Variables in Accurate Forecasts are Dynamic
- Relevant Variables are Difficult to Discern
- Adversaries may React to Accurate Forecasts
- - Forecasted Outcomes may be Prevented
- Unforeseen Actors will Always Emerge

Although intelligence provides a great advantage in supporting organizational interests, there are significant limitations inherent to it as well.

Chapter 01 Summary: Key Concepts

- Describing Intelligence by Type
- The Primary Mission: Decision Advantage
- Common Analytical Errors
- Intelligence Sources and Methods
- Strategic, Operational, and Tactical Intelligence
- The Limitations of Intelligence

PHASE2 ADVANTAGE © 2023 CHAPTER 01

The Intelligence Process and Life Cycle

PHASE2 ADVANTAGE © 2023 CHAPTER 02

The Intelligence Process (I)

- A Large Volume of Information is Available
- - Most is Misleading or Irrelevant
- - Must be Examined and Discarded
- Incoming Information Saturates Analysts
- Critical Data may not be Shared Effectively

Although most of the focus is on collection and operations, most of the failures in intelligence are due to inadequate or nonexistent analysis.

PHASE2 ADVANTAGE © 2023 CHAPTER 02

The Intelligence Process (II)

- Intelligence is Always Concerned with Targets
- - The focus of the Problem Needing Answers
- **Goal**: *How to Approach the Target*
- - Steps to Solving the Intelligence Problem
- - Using a Target-Centric Approach
- - Communicate Understanding to Customers
- - Customers can Act on the Understanding

The analyst's primary job is to develop a level of understanding of the target and communicate that knowledge to the customer of the intelligence product.

Analysis Scenario (Rules of Engagement)

ROE 1:

No actions may be taken to alter the environment or any of the components presented in this scenario. Nothing may be added, removed, or manipulated by participants in this analysis. All facts have been presented; there is nothing "hidden." However, there may be questions needed for clarification.

ROE 2:

Participants will have 5 minutes to read the scenario, conduct their analysis, and be prepared to submit their answer based on the Issue Decomposition.

ROE 3:

When you have determined your answer, write it down and notate your justification. Do not discuss the answer until the end of the exercise.

ROE:
Rules of Engagement

A Risky Jump for the Family

Situation:

Who would jump out of an aircraft without a parachute? If **you** agree to take the risk, you will be paid $10 million USD – **if** you survive. If you don't, the $10 million USD is **guaranteed** to be paid to your family, ensuring financial stability for the rest of their lives.

An experienced skydiving instructor will jump with you and will have an extra parachute (for **you**, and it works **perfectly**). You can ask for it **any** time before or during the jump. The instructor will give it to you and secure it properly **without hesitation**.

However, if you ask for (and accept) the parachute, you – and your family – will **immediately** lose the opportunity to receive the $10 million USD. On the bright side, both you and your family will remain safe and healthy.

IR Tasking Order:

Based on the facts, would you be willing to selflessly risk your health and well-being to guarantee the future financial stability of you and your family?

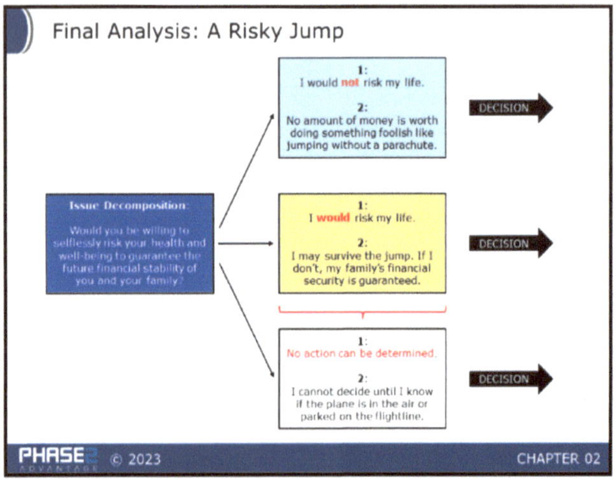
Final Analysis: A Risky Jump
Issue Decomposition:
Would you be willing to selflessly risk your health and well-being to guarantee the future financial stability of you and your family?
1:
I would not risk my life.
2:
No amount of money is worth doing something foolish like jumping without a parachute.
DECISION
1:
I would risk my life.
2:
I may survive the jump. If I don't, my family's financial security is guaranteed.
DECISION
1:
No action can be determined.
2:
I cannot decide until I know if the plane is in the air or parked on the flightline.
DECISION
PHASE2 ADVANTAGE
© 2023
CHAPTER 02

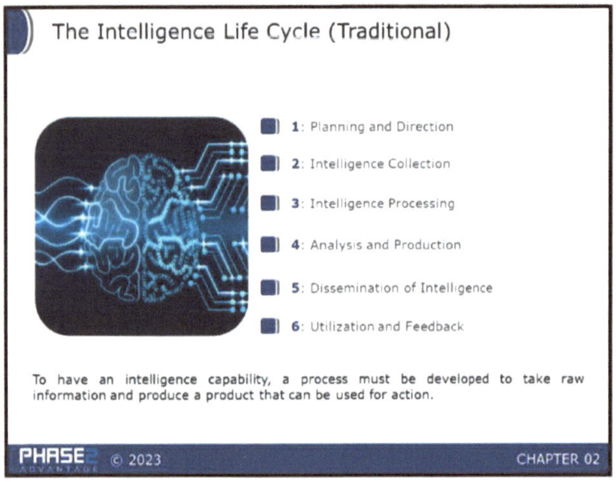
The Intelligence Life Cycle (Traditional)
1: Planning and Direction
2: Intelligence Collection
3: Intelligence Processing
4: Analysis and Production
5: Dissemination of Intelligence
6: Utilization and Feedback
To have an intelligence capability, a process must be developed to take raw information and produce a product that can be used for action.
PHASE2 ADVANTAGE
© 2023
CHAPTER 02

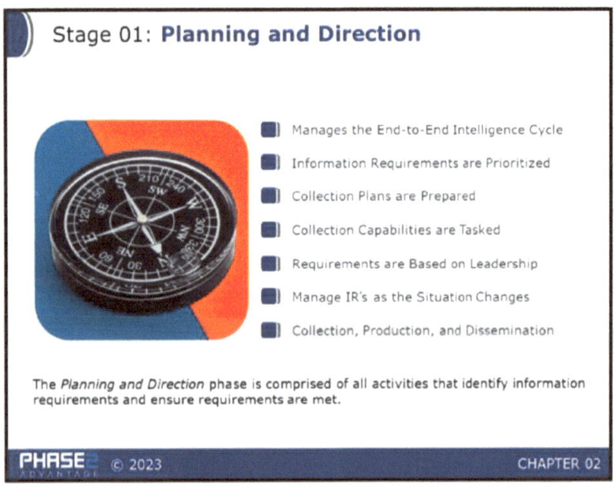
Stage 01: Planning and Direction
Manages the End-to-End Intelligence Cycle
Information Requirements are Prioritized
Collection Plans are Prepared
Collection Capabilities are Tasked
Requirements are Based on Leadership
Manage IR's as the Situation Changes
Collection, Production, and Dissemination
The Planning and Direction phase is comprised of all activities that identify information requirements and ensure requirements are met.
PHASE2 ADVANTAGE
© 2023
CHAPTER 02

Stage 02: **Intelligence Collection**

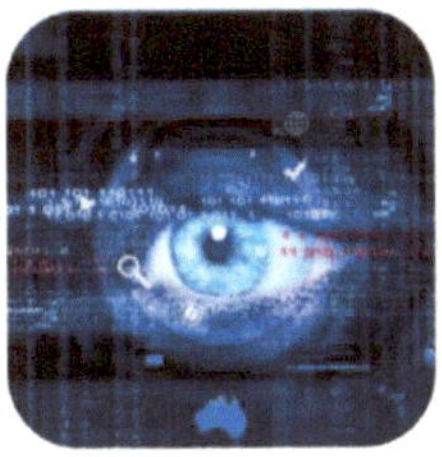

- Allow Teams to Collect a Portion of IR
- Contributes to Information Redundancy
- - Information to Confirm Assessments
- - Information to Disprove Assessments
- Facilitates Cross-Communication
- Collected Information is Reconciled
- Collected Information is Forwarded

The *Intelligence Collection* phase acquires and provisions information to be provided to teams in the next chain of the intelligence cycle.

PHASE2 ADVANTAGE © 2023 CHAPTER 02

Stage 03: **Intelligence Processing**

- Process and Exploitation are Conducted
- - Conversion Includes **Decryption**
- - Conversion Includes **Translation**
- - Conversion Includes **Interpretation**
- Now Regarded as Actual Intelligence
- Uses a Format Specific to the Organization

During the *Processing* phase, processed information regarded as actual intelligence will be mapped to correlating points for future phases.

PHASE2 ADVANTAGE © 2023 CHAPTER 02

Stage 04: **Analysis and Production**

- Products Focused on Stakeholder Need
- Delivered in a Timely Manner
- - Intelligence Report is Complete
- - Intelligence Report is Accurate
- Determines Additional Collection Efforts
- Provides an Overview of the Subject Area
- Allows the Stakeholder to Draw Conclusions

The *Analysis and Production* phase analyzes, evaluates, interprets, and integrates raw intelligence into finished intelligence products.

PHASE2 ADVANTAGE © 2023 CHAPTER 02

Stage 05: **Dissemination of Intelligence**

- Many Architectures Exist for Dissemination
- - *Supply-Push*: No Request Needed
- - *Demand-Pull*: As Needed by Querying
- - *Standard Channel*: Routine Intelligence
- - *Alarm Channel*: Critical Intelligence
- - *Broadcast Mode*: All Authorized Users
- - *Point-to-Point Mode*: Specific Users

Intelligence Dissemination portrays an accurate picture of the threat to leadership in a form that facilitates rapid understanding of the issue.

Stage 06: **Utilization and Feedback**

- The Intelligence Life Cycle is Closed-Loop
- Feedback Between Phases is Fluid
- Consumers are not Required to Act
- *Are IR's Adequately Met by Analysts?*
- - Analysts Address Needed Adjustments
- Revisions / Requirements to IR are Issued
- Determine if More Collection is Needed

The ***Utilization and Feedback*** phase ensures a dialog occurs between all parties before and after the intelligence has been received.

Challenges to the Traditional Life Cycle Model

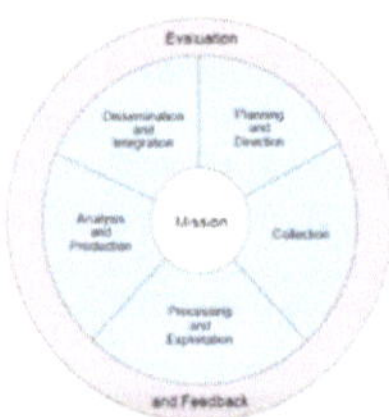

- Effective Intelligence Efforts are not Cycles
- Defines an Antisocial Series of Steps
- Separates the Analysts from Collectors
- Responsibility for the Final Product is Avoided
- Predictable / Vulnerable to Countermeasures
- Treats the Customer as an Abstract Concept
- True Cycle Feedback Loops are Absent

The traditional model may adequately describe the structure and function of the intelligence *community*, but it does not describe the intelligence *process*.

The Intelligence Life Cycle (Target-Centric)

- **A Network-Centric Collaboration Process**
- **Goal**: Construct a Shared Target Picture
- Process can Handle Complex Problems
- Takes Advantage of IT Implementations
- All Participants can Contribute Knowledge
- Not a Linear Sequence or Traditional Cycle
- It is a **Network and Social Process**

An alternative to the traditional intelligence cycle is to make all stakeholders, including customers, an intrinsic part of the intelligence process.

The Team-Generated Intelligence Model

- Intended to Encourage Interaction
- More Interactive than the Traditional Cycle
- A Better Way to Handle Complex Problems
- Participants Share Knowledge of the Target
- - Better able to Share Gaps in Knowledge
- - Better able to Understand Critical Issues
- **All Resources Focused on the Target**

In the target-centric process, collectors and customers can not only see into the process but have many more opportunities to contribute to it.

The Target as a Network

- Systems are Viewed as Individual Targets
- - Narrow Focus Downplays Connections
- - Connections can be Real Strengths
- - Connections can be Real Weaknesses
- Networks Comprised of **Nodes** and **Links**
- - People, Places, Things, or Concepts
- - Links Defines the Relationship of Nodes

The goal of intelligence is to develop knowledge of the adversary network, make the analyst's network effective, and to render the adversary network ineffective.

Components of an Effective Intelligence Network

- Robust Communications Connectivity
- Leverages Physical and Cultural Proximity
- Network Members have a Shared Purpose
- Established Decision-Making Processes
- Professional and Personal Relationships
- Members Effectively Decide and Act
- A Flexible / Empowered Network is Difficult

It may be easier to see the adversary's side as a network than to see that one's own intelligence assets form a network and to fully exploit its strengths.

PHASE2 ADVANTAGE © 2023 CHAPTER 02

Object-Based Production (Target-Centric)

- Organizes Efforts Around "Objects" (Targets)
- Features Cloud-Based Sharing of Knowledge
- The Object is a Bucket to Store Information
- - Becomes a Single Point of Convergence
- - Serves as a Launching Point for Intelligence
- OBP is an Intentional Process Method

Object-based production is being implemented as a whole-of-community concept that fundamentally changes the way information and intelligence is organized.

PHASE2 ADVANTAGE © 2023 CHAPTER 02

Spatial and Temporal Target Attributes

- Many Targets are Fixed **Geographically**
- - Elements of a Region's Infrastructure
- - Data Centers, Bandwidth, Power Grid
- Many Targets are Mobile (**Intermittently**)
- - Characterized by both Space and Time
- Techniques to Locate / Track Internet Users
- - Geolocating IP's / Hardwired Features

In addition to being a system and a network, targets of intelligence interest have spatial attributes and temporal attributes that must be considered by analysts.

PHASE2 ADVANTAGE © 2023 CHAPTER 02

Benefits of Target-Centric Intelligence

- Analysis can take Many Forms
- Analysis can Draw on Many Sources
- Several Types of Sources are Utilized
- Multiple Sources Creates a Complete Picture
- Extrinsic Sources are Often Utilized
- Analysis is Always a Continuous Process
- Process Functions are Always Ongoing

The analyst's job is to take information from disparate sources and provide support to many customers with different requirements and needs.

Chapter 02 Summary: Key Concepts

- The Intelligence Process
- The Intelligence Life Cycle (Traditional)
- The Intelligence Life Cycle (Target-Centric)
- The Target as a Network
- The Team-Generated Intelligence Model
- Benefits of Target-Centric Intelligence

Adversaries in the Modern Threat Landscape

Cyber Adversaries vs. Cyber Defenders

Adversaries	Defenders

No Regard for Laws	Hack Back Prohibited
No Limits or Boundaries	Regulation and Compliance
Tax Free Funding	Limited Budgets and Personnel
Acquire Exploit Technology	Change and Version Control
Unlimited Time Frame	Prove Security Value Daily
Must be Right Once	Must be Right Always

Nation-State Actors

- **Sovereignty**: Authority to Rule
- **Territoriality**: Recognized Borders
- **Abstraction**: Independent Persona
- Exert Influence Over Citizens
- Indirect Economic Support to Actors
- Sony Pictures Headquarters (2014)
- Difficult to Identify a Policy Response

The role of *nation-states* in cyberattacks that involve hacking and data theft has gained substantial attention over the last two decades.

Advanced Persistent Threats (APTs)

- Motivations are Political and Economic
- Targeted Sectors Include all Industries
- APTs are Highly Skilled and Knowledgeable
- - Threat Actors have Access to Intelligence
- - Threat Actors have Specific Objectives
- - Threat Actors have Capability and Intent
- Multiple Targeting Methods often Combined
- Executed by Coordinated Human Actions

An *Advanced Persistent Threat* (APT) is a threat actor, typically a nation-state or state-sponsored group, attempting to gain unauthorized access to a network.

8 Unique Characteristics of APTs

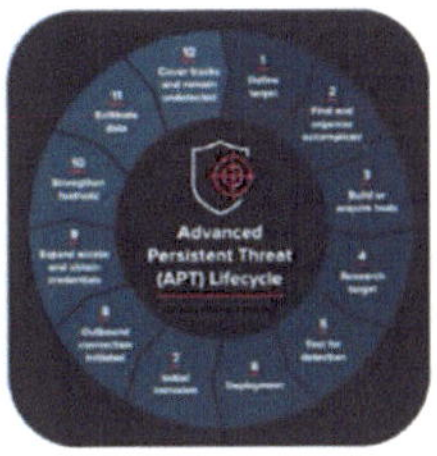

- **1**: Actors
- **2**: Objectives and Methods
- **3**: Resources
- **4**: Time of Operation
- **5**: Resources
- **6**: Risk Tolerance
- **7**: Origin of Attacks
- **8**: Value of Attacks

Many common attack vectors were initially introduced as part of an APT campaign using zero-day exploits, malware, and customized credential theft.

Examples of APTs Threatening U.S. Interests

- **APT38** (aka: *Lazarus Group*)
- - Attribution: North Korea
- **APT31** (aka: *Judgment Panda*)
- - Attribution: China
- **APT35** (aka: *Newscaster Team*)
- - Attribution: Iran
- **APT33** (aka: *Elfin Team*)
- - Attribution: Iran

Traditional cybersecurity measures such as defense-in-depth, firewalls, and antivirus cannot protect against a sophisticated and well-planned APT attack.

Non-State Actors

- Tend to have Fewer Resources
- Attacks Tend to be Less Sophisticated
- May have Different Methods and Targets
- Can Directly Profit from Data Theft
- Can Affect Operational Capabilities
- Can Cause Fear Among the Population
- Not Restricted by Command Hierarchies

Non-state actors may work individually or complete attacks with a wide network of participants and then disband rapidly.

Components of the Threat Landscape

- Economic Crimes and Identity Theft
- Intellectual Property Theft and Piracy
- Cyber-Stalking and Online Harassment
- Hacktivism and Online Extremism
- Cyber-Terror and Cyber-Warfare
- Computer Hackers and Hacking

The Internet and computer technologies have dramatically reshaped the threats posed to modern society and the ways we defend against them.

Economic Crimes and Identity Theft

- Email is a Critical Resource for Attackers
- A Time-Sensitive Problem is Stressed
- PII Stolen for Fraud or Impersonation
- Healthcare Data Storage is Tremendous
- Attackers Target for Mass Exploitation
- Emergence of Darknet Carding Markets
- Majority Report to Banks (Not Police)

The ability to buy and sell goods at any time or location is revolutionary for both commerce and crime.

Intellectual Property Theft and Piracy

- Copyright, Trademark, and Patent
- Trade Secrets for Strategy or Processes
- Low-Income Countries: Software Piracy
- Continuously Changing Behaviors
- Bit-Torrent Links to Seeder Servers
- - Allows Concurrent Uploads / Downloads
- - Reduces Risk and Likelihood of Detection

Ideas become *Intellectual Property* when they are physical and tangible works that may be accessed and viewed by others.

Cyber-Stalking and Online Harassment

- Technology can Cause Real-World Harm
- Involves Constant Use of Email and Text
- Ability to Gather Personal Information
- Victims Report Feeling Powerless
- - Anxiety, Depression, and Absenteeism
- Exposure Reduced by Changing Behavior
- Does not Need to Know the Victim's Identity

The open nature in which people can lead their lives is unparalleled—except by the access provided to their personal information.

PHASE2 ADVANTAGE © 2023 CHAPTER 03

Hacktivism and Online Extremism

- Technology to Promote an Activist Agenda
- Motive of Actor vs. People Harmed
- Tendency to Target Larger Groups
- Designed to Target Innocent People
- - Can Cause General Panic and Fear
- - Can Cause Mass Collateral Damage
- Attacks may Rival Physical Incidents

A carefully targeted attack against any critical infrastructure asset can cause serious harm to the functionality and security of the network.

PHASE2 ADVANTAGE © 2023 CHAPTER 03

Hacktivism in Action: ***Anonymous***

- The Collective Originated in 2003
- Members are Known as "**ANONS**"
- Associated with Online Hacktivism
- Operates Domestically and Internationally
- Motivated by Issues and Ideologies
- No Definitive Leadership Exists
- - Attribution for Group Actions is Difficult
- Engages in Direct Physical Action Offline

Anonymous is a decentralized and international hacktivist group known for its cyber-attacks against governments, institutions, agencies, and corporations.

PHASE2 ADVANTAGE © 2023 CHAPTER 03

Cyber-Terror and Cyber-Warfare

- No Agreed Definition for Cyber-Warfare
- Twitter is a Key Platform for Radicalization
- Al Qaeda Operates the Magazine *Inspire*
- - Detailed Tactics to Conduct Terrorism
- eJihad: *al-Farouq* and *al-Jinan* Forums
- - Ability to Obtain Free Exploit Tools
- USA: *Ku Klux Klan* and *Aryan Nations*

The ability to communicate with others of diverse backgrounds ensures that individuals can be steadily introduced to a movement's principles.

PHASE ADVANTAGE © 2023 CHAPTER 03

Computer Hackers and Hacking

- Many Use Non-Technical Strategies
- Humans are the Greatest Vulnerability
- The Technology Used is Ethically Neutral
- Black Hat, Gray Hat, and White Hat Hackers
- Online Relationships can Reinforce Behavior
- Much Time is Devoted to Study and Practice

A *hack* involves the modification of technology in order to allow it to be used in unintended ways for both legal and illegal purposes.

PHASE ADVANTAGE © 2023 CHAPTER 03

Script Kiddies (aka: "Skiddies" or "Skids")

- Uses Existing and Proven Techniques
- Uses Existing Programs and Scripts
- Motivated by Basic and Personal Reasons
- - Fun, Chaos, Attention, or Revenge
- Tend to be Less Experienced with Exploits
- - Possesses Less Developed Hacking Skills
- Little Regard for Potential Consequences
- Still Considered to be a Serious Threat

A *Script Kiddie* is a person who is not technologically sophisticated, who randomly seeks out a specific weakness to gain access to a network or system.

PHASE ADVANTAGE © 2023 CHAPTER 03

Script Kiddies: An Underestimated Threat

- Social Engineering is Used Extensively
- - Phishing, Vishing, and Fake Websites
- Can Cause Extreme Damage and Disruption
- Actions Tend to be Unpredictable
- - Skilled Hackers Rely on Planned Exploits
- May Unwittingly Aid Serious Criminals
- - Sophisticated Actors can Remain Hidden
- Script Kiddies are **Cyber Criminals**

A *Script Kiddie* can easily find vulnerabilities to access private systems using programs intended for computer security and attack networks and websites.

Legal Challenges in Intelligence Collection

- Electronic Communications Privacy Act
- The Fourth Amendment
- The Fifth Amendment
- International Jurisdictions
- *Nulla Poena Sine Lege*

Evolving privacy and data protection regulations across geographies may add to the complexity of gathering actionable intelligence and accessing sources.

Electronic Communications Privacy Act

- Amendments to *Wiretap Act*
- Updates to *Stored Communications Act*
- Created the *Pen Register Act*
- Bars Electronic Eavesdropping
- Disclosure of Illegally Obtained Data
- Unlawfully Accessing Data at Rest
- Prohibits Pen Register and Trace Devices

The *Electronic Communications Privacy Act* (ECPA) was passed in 1986 to expand and revise wiretapping and electronic eavesdropping provisions.

The Fourth Amendment

- Only Applies to Government Agencies
- A Reasonable Expectation of Privacy
- Exigent Circumstances in Emergencies
- The *Apparent Authority* Principle
- Shared Authority Through Relationships
- Courts: Phones Hold the Privacies of Life
- Consent is the Most Relevant Exception

The *Fourth Amendment* limits the legal authority of government—or its agents—to search and seize evidence without a warrant.

PHASE2 © 2023 CHAPTER 03

The Fifth Amendment

- Protection Against Self-Incrimination
- Silence does not Indicate Guilt
- Access Codes and Device Encryption
- Compelling Violates the Amendment
- Thumbprints and Facial Recognition
- No Key Disclosure Laws in the US
- Consent is the Most Relevant Exception

The *Fifth Amendment* protects people from self-incrimination, meaning giving a statement that might implicate oneself in a crime.

PHASE2 © 2023 CHAPTER 03

Nulla Poena Sine Lege

- A Basic Requirement for the Rule of Law
- *Nulla Poena Sine Lege* ***Praevia*** (Prior)
- *Nulla Poena Sine Lege* ***Scripta*** (Written)
- *Nulla Poena Sine Lege* ***Certa*** (Clear)
- *Nulla Poena Sine Lege* ***Stricta*** (Exact)

Nulla poena sine lege is a universal legal principle, requiring that one cannot be punished for doing something that is not prohibited by law.

PHASE2 © 2023 CHAPTER 03

Respected Threat Intelligence Sources

- AlienVault Open Threat Exchange
- Center for Internet Security
- CIA World Factbook
- Cisco GOSINT
- FBI InfraGard
- Mandiant (Fireye)
- MITRE ATT&CK
- SANS Internet Storm Center

Threat intelligence can be derived from external sources such as open-source platforms, or communications between threat information sharing groups.

PHASE2 ADVANTAGE © 2023 CHAPTER 03

Chapter 03 Summary: Key Concepts

- Cyber Adversaries vs. Cyber Defenders
- Nation-State Threat Actors
- Advanced Persistent Threats
- Nonstate Threat Actors
- Legal Challenges in Intelligence Collection
- Respected Threat Intelligence Sources

PHASE2 ADVANTAGE © 2023 CHAPTER 03

Enhancing Analytical and Logical Reasoning Skills

PHASE2 ADVANTAGE © 2023 CHAPTER 04

Overview of Basic Thinking Abilities

- Basic Abilities and Complex Skills
- Relate to the Analyst's Ability to Think
- Continually Improve the Quality of Thinking
- Three Basic Thinking Abilities are Required:
- **- Information Ordering**
- **- Pattern Recognition**
- **- Logical Reasoning**

When communicating the results of analysis, the analyst must ensure that the customer understands the conclusions and how they were reached.

PHASE2 ADVANTAGE © 2023 CHAPTER 04

Information Ordering

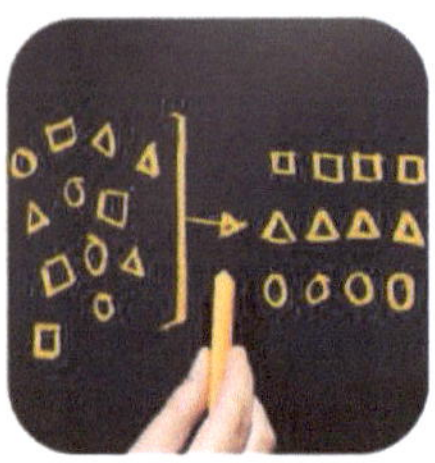

- Allows Analysts to Arrange Information
- Often with the Assistance of Technology
- Analysis, Synthesis, and Higher Understanding
- Leads the Analyst to Make Conclusions
- Disseminate Conclusions as Intelligence
- Ordering is Inherently Limiting
- - May not See Alternative Explanations

Information ordering is the ability to follow predefined rules or sets of rules to arrange the data in a meaningful order that enhances the intelligence process.

PHASE2 ADVANTAGE © 2023 CHAPTER 04

Pattern Recognition

- Impose Patterns to Detect Relationships
- Often Infers Potential Actions in the Future
- Conceptualize a Degree of Order from Chaos
- Separate Relevant from Irrelevant Facts
- **Risk**: Seeking Patterns can Introduce Bias
- - May Impose Culturally-Defined Patterns
- - May not Recognize Inherent Patterns

Analysts detect and impose *patterns* on apparently random entities and events in order to understand them, often doing this without being aware of it.

PHASE2 ADVANTAGE © 2023 CHAPTER 04

Indicators of Falsified Information (I)

EMPLOYEE "A"		EMPLOYEE "B"	
Employer Information		**Employer Information**	
Employer:	Geek Tech Computer Repair	Employer:	Savannah Bytes Programming
Address:	438 Mooreland Avenue NE	Address:	124 Stevens Court SW
City, State, Zip Code:	Marietta, Georgia 30062	City, State, Zip Code:	Savannah, Georgia 30124
Telephone:	(770) 794-9390	Telephone:	(912) 487-5050
Applicant Information		**Applicant Information**	
First Name:	Nicholas	First Name:	Michael
Last Name:	Ferron	Last Name:	Guifanno
Address:	829 Jefferson Street NW	Address:	1 Riverside Drive, Apt. 24a
City, State, Zip Code:	Atlanta, Georgia 30308	City, State, Zip Code:	Rincon, Georgia 31247
Telephone:	(404) 506-1923	Telephone:	(912) 493-3446
Date of Birth:	11/15/1977	Date of Birth:	01/24/1978
Social (SSN):	157-43-1560	Social (SSN):	250-12-4693
Driver License:	(GA) 047829025	Driver License:	(GA) 030612477

The applicant information displayed in this example is fictitious and was created for the purpose of this pattern recognition exercise.

 CHAPTER 04

Indicators of Falsified Information (II)

EMPLOYEE "A"		EMPLOYEE "B"	
Employer Information		**Employer Information**	
Employer:	Geek Tech Computer Repair	Employer:	Savannah Bytes Programming
Address:	438 Mooreland Avenue NE	Address:	124 Stevens Court SW
City, State, Zip Code:	Marietta, Georgia 30062	City, State, Zip Code:	Savannah, Georgia 30124
Telephone:	(770) 794-9390	Telephone:	(912) 487-5050
Applicant Information		**Applicant Information**	
First Name:	Nicholas	First Name:	Michael
Last Name:	Ferron	Last Name:	Guifanno
Address:	829 Jefferson Street NW	Address:	1 Riverside Drive, Apt. 24a
City, State, Zip Code:	Atlanta, Georgia 30308	City, State, Zip Code:	Rincon, Georgia 31247
Telephone:	(404) 506-1923	Telephone:	(912) 493-3446
Date of Birth:	11/15/1977	Date of Birth:	01/24/1978
Social (SSN):	157-43-1560	Social (SSN):	250-12-4693
Driver License:	(GA) 047829025	Driver License:	(GA) 030612477

The applicant information displayed in this example is fictitious and was created for the purpose of this pattern recognition exercise.

 CHAPTER 04

Logical Reasoning

- Transform Information into Intelligence
- Four Types of Reasoning Guide Analysts
- **- Inductive Reasoning**
- **- Analogical Reasoning**
- **- Deductive Reasoning**
- **- Abductive Reasoning**

The ability to *reason* is what permits analysts to process information and formulate explanations, to assign meaning to observed actions and events.

 CHAPTER 04

Inductive Reasoning

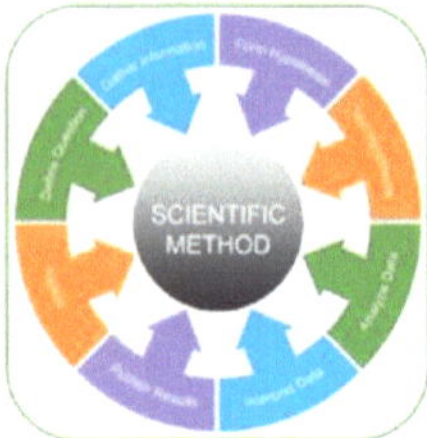

- Induction Normally Precedes Deduction
- Most Common Type of Analytical Reasoning
- - Requires **Objectivity**
- - Requires **Elimination of Prejudice**
- - Requires **Elimination of Preconceptions**
- Can be a New Source of Knowledge
- Dependent Upon Accurate Observation

Inductive reasoning is a process of discovery in which an analyst establishes a relationship between events under observation or study.

Analogical Reasoning

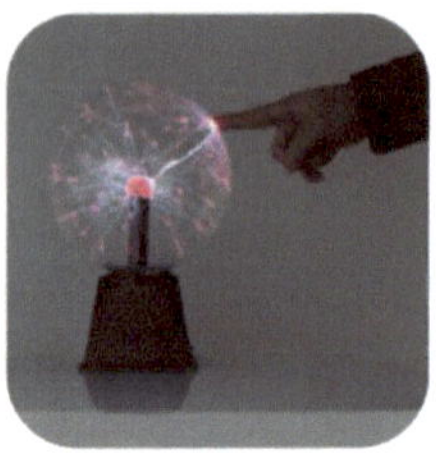

- Similarities are Used to Gain Understanding
- - Substantial Similarities Between Situations
- - Likely to be Further Similarities
- Facts are Known about a Situation
- Facts can be Used to Infer New Information

Analogical reasoning is a method of processing information that compares the similarities between new concepts and previously understood concepts.

Deductive Reasoning

- Analysts Begin with a Set of Rules
- Rules are Used to Interpret Information
- Argument is Sound if the Premise is True
- **Risk**: May Lead to an Incorrect Conclusion
- - A Series of Events Precede an Action
- - A Mistake in Confusing *Cause and Effect*
- Not Effective Forecasting Human Behavior

Deductive reasoning is a logical process in which a conclusion is based on the concordance of multiple premises that are generally assumed to be true.

Abductive Reasoning

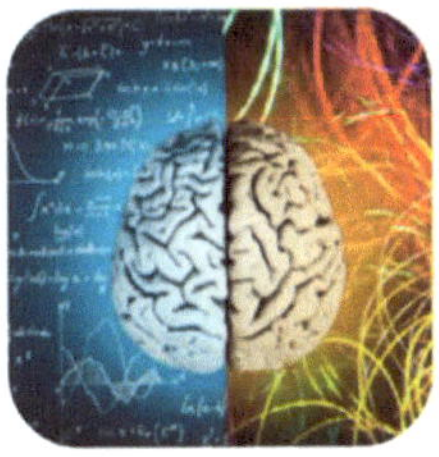

- Available Information ≠ Expected Outcome
- The Reason must be Determined
- A New Hypothesis must be Generated
- - Why Given Evidence ≠ Familiar Explanation
- **Disassociate**, **Re-Order**, then **Re-Engage**
- Test the New Explanation for Validity

Abductive reasoning describes the logical thought process that accompanies the insight, intuition, or experience of the analyst conducting the research.

Abductive Reasoning: Example and Exercise

Available Information:

A tennis racket and a tennis ball cost a total of one dollar and ten cents.
The tennis racket costs one dollar more than the tennis ball.

Question:

How much does the tennis ball cost?

Critical Thinking: The 8 Elements of Thought

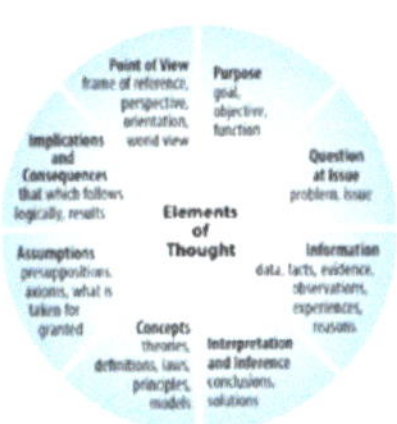

- **1**: Purpose and Goals
- **2**: Question at Issue
- **3**: Available Information
- **4**: Interpretation and Inference
- **5**: Accepted Concepts
- **6**: Assumptions and Opinions
- **7**: Implications and Consequences
- **8**: Point of View

Analysts often tend to believe their thoughts are accurate just because they are theirs; therefore, the thoughts of those that disagree with them are inaccurate.

The 8 Intellectual Standards of Analysis (I)

- A Solid Foundation for Critical Thinking
- Critical Thinking is a Deliberate Choice
- Requires Focus and Self-Discipline
- Requires Commitment to Improve Skills
- Ensures Analysts are Effective and Efficient
- - The Most Complicated or Ambiguous Tasks

While the *elements of thought* provide a framework for **analyzing** thinking, the *standards of thought* provide criteria to assess the **quality** of thinking.

PHASE2 ADVANTAGE © 2023 CHAPTER 04

The 8 Intellectual Standards of Analysis (II)

- **1**: Fair-Mindedness
- **2**: Intellectual Humility
- **3**: Intellectual Courage
- **4**: Intellectual Empathy
- **5**: Intellectual Integrity
- **6**: Intellectual Perseverance
- **7**: Confidence in Reason
- **8**: Intellectual Autonomy

When analysts dissect their thinking and examine its parts, they use standards of quality referred to as *intellectual standards* or *standards of thought*.

PHASE2 ADVANTAGE © 2023 CHAPTER 04

Avoiding Analytical Pitfalls

- Analysis is Subject to Numerous Influences
- Analysts are Prone to Numerous Influences
- - Shape and Mold View of the World
- - Shape and Mold the Ability to Reason
- These Influences are **Analytical Pitfalls**
- Pitfall 01: **Biases**
- Pitfall 02: **Logical Fallacies**

Elements of thought, intelligence standards, and intellectual traits aid analysts in recognizing pitfalls in their own analysis and the analysis performed by others.

PHASE2 ADVANTAGE © 2023 CHAPTER 04

Biases (I)

- **Cultural Bias**
- - Referred to as *"Mirror Imaging"*
- **Organizational Bias**
- - Group Think
- - Best Case
- **Personal Bias**
- - Assimilation Bias

A *bias* is a disproportionate weight in favor of or against an idea or thing, usually in a way that is closed-minded, prejudicial, or unsupported by fact.

Biases (II)

- - Confirmation Bias
- - Anchoring Bias
- **Cognitive Bias**
- - Vividness
- - Oversensitivity to Consistency
- - Dependency on Memory
- - Acceptance of New Intelligence

Biases generally have a detrimental impact on intelligence analysis because they obscure the true nature of the information (both received and disseminated).

Logical Fallacies of **Relevance** (I)

- **Appeal to Force**
- - Last Resort Intimidation when Logic Fails
- **Genetic Fallacy**
- - An Idea is Wrong Based on its Origin
- **Argument to the Man**
- - Discredit the Source to Win the Argument
- **Fallacy of Hasty Generalizations**

Although usually committed unintentionally, these fallacies are sometimes used deliberately to persuade, convince, or deceive those unaware that they exist.

Logical Fallacies of **Relevance** (II)

- **Fallacy of Division**
- - The Whole Represents all its Parts
- **Post Hoc Ergo Propter Hoc**
- - Distorting Reasons for Cause and Effect
- **Fallacy of False Dilemma**
- - Referred to as *"Black and White Thinking"*
- **Fallacy of Special Pleading**

An analyst must be able to recognize logical fallacies so a false line of reasoning will not distract them and lead to poor, unsupported conclusions.

PHASE2 © 2023 CHAPTER 04

Logical Fallacies of **Omission**

- **Oversimplification**
- - Generality Fails to Account for Complexity
- **Fallacy of Composition**
- - Features of Parts Represent the Whole
- **Misusing Analogies**
- - Indiscriminate Generalizing of Analogies

Fallacies of omission occur when an analyst leaves out necessary material (either intentionally or unintentionally) in a conclusion or inference.

PHASE2 © 2023 CHAPTER 04

Chapter 04 Summary: Key Concepts

- Overview of Basic Thinking Abilities
- Critical Thinking: The Elements of Thought
- The 8 Intellectual Standards of Analysis
- Avoiding Analytical Pitfalls
- An Overview of Biases
- An Overview of Logical Fallacies

PHASE2 © 2023 CHAPTER 04

Managing the Information Collection Process

PHASE2 ADVANTAGE © 2023 CHAPTER 05

The Role of the Collection Manager (I)

- The Collection Manager Plays a Critical Role
- - Communicates Analyst Needs to Collectors
- - Generates Collection Requirements
- - Responds to Intelligence Gaps
- - Understands the Collector's Capabilities
- - Understands the Analyst's Needs
- - Selects Appropriate Intelligence Disciplines

In the field of intelligence, the facilitator between the producers of intelligence (analysts) and the operators (collectors) is the *Collection Manager*.

PHASE2 ADVANTAGE © 2023 CHAPTER 05

The Role of the Collection Manager (II)

- Facilitates Feedback to Collectors
- Ensures Collected Information Meets Needs
- - Additional Information may be Required
- - Analysts may have Additional Questions
- Helps Collectors Improve the Process

The *Collection Manager* ensures and maintains an efficient two-way line of communication always exists between the analysts and the collectors.

PHASE2 ADVANTAGE © 2023 CHAPTER 05

The Role of the Collection Manager (III)

- Data Mining and Research
- Collaboration in Real Time
- Conducting Link and Pattern Analysis
- Developing Threat and Warning Templates
- Insulating the Team from External Forces
- Maintaining Intelligence Databases and Files
- Disseminating Information Amongst Teams

The *Collection Manager* operates across all echelons, all security and network domains, and across all computing environments to aid intelligence personnel.

Analysis Scenario (Rules of Engagement)

ROE 1:

No actions may be taken to alter the environment or any of the components presented in this scenario. Nothing may be added, removed, or manipulated by participants in this analysis. All facts have been presented; there is nothing "hidden." However, there may be questions needed for clarification.

ROE 2:

Participants will have 5 minutes to read the scenario, conduct their analysis, and be prepared to submit their answer based on the Issue Decomposition.

ROE 3:

If you believe you have solved the IR Tasking Order, write it down for the end of the exercise. Do not discuss the answer until the exercise is over. Be prepared to discuss the reasoning that led to your answer.

The Air Force Generals Disagree

Situation:

An additional 500 pounds of armoring has been authorized for a surveillance aircraft. One General wants to armor the *avionics* (electronic systems) in the nose of the aircraft, the other General wants to armor the *fuselage* (body) of the aircraft. **Only one location can be chosen**.

Your team has been sent to a base to analyze the situation. Every day you observe **10** surveillance aircraft leave the base on missions. **80%** of the aircraft return to the base after completing the mission. It is reported by returning pilots those that did not return were **shot down** during the mission.

You notice that **100%** of the returning aircraft have a significant number of bullet holes in the fuselage an **no** damage to the avionics. This continues for **every** mission. You document the results of **10 missions** and are asked to advise the Generals where the armor must be placed to protect the planes.

IR Tasking Order:

Based on your analysis, where would you recommend placing the additional armor on the aircraft? The **avionics**, or the **fuselage**?

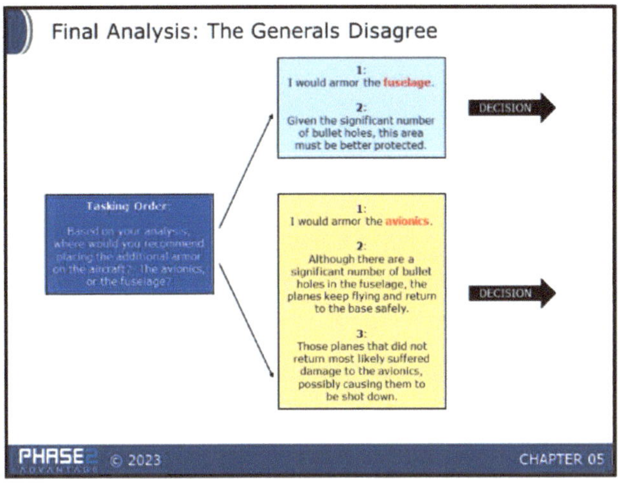

Final Analysis: The Generals Disagree
Tasking Order:
Based on your analysis, where would you recommend placing the additional armor on the aircraft? The avionics, or the fuselage?
1:
I would armor the fuselage.
2:
Given the significant number of bullet holes, this area must be better protected.
DECISION
1:
I would armor the avionics.
2:
Although there are a significant number of bullet holes in the fuselage, the planes keep flying and return to the base safely.
3:
Those planes that did not return most likely suffered damage to the avionics, possibly causing them to be shot down.
DECISION
PHASE2 ADVANTAGE © 2023
CHAPTER 05

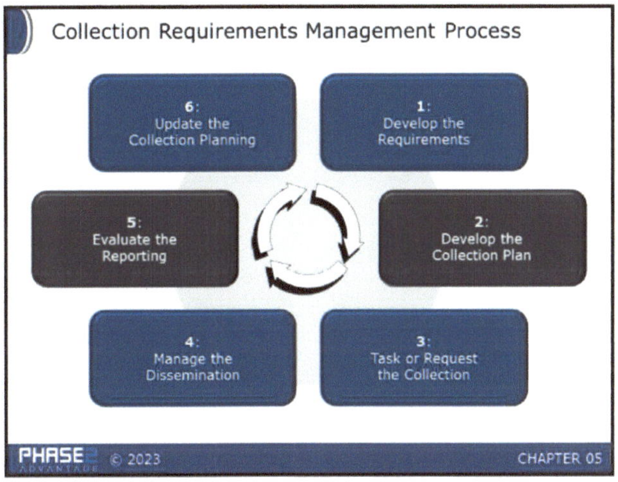

Collection Requirements Management Process
6:
Update the Collection Planning
1:
Develop the Requirements
5:
Evaluate the Reporting
2:
Develop the Collection Plan
4:
Manage the Dissemination
3:
Task or Request the Collection
PHASE2 ADVANTAGE © 2023
CHAPTER 05

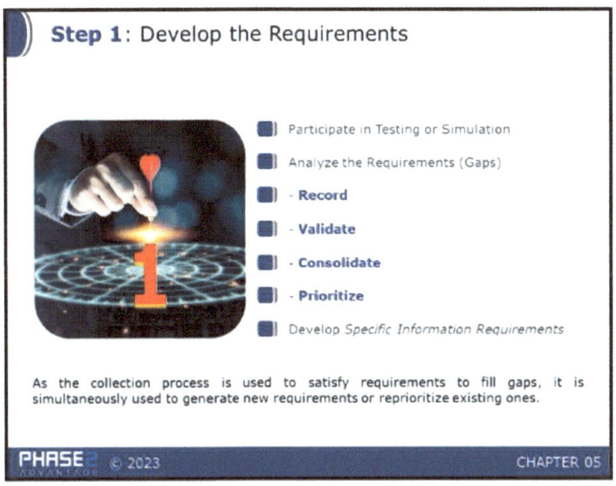

Step 1: Develop the Requirements
Participate in Testing or Simulation
Analyze the Requirements (Gaps)
- Record
- Validate
- Consolidate
- Prioritize
Develop Specific Information Requirements
As the collection process is used to satisfy requirements to fill gaps, it is simultaneously used to generate new requirements or reprioritize existing ones.
PHASE2 ADVANTAGE © 2023
CHAPTER 05

Conducting a Gap Analysis

- **The Gap Analysis Process**
- - Prioritize Gaps Based on Importance
- - Prioritize Gaps Based on Size
- Classify Gaps Based on their Nature
- - Collection, Processing, or Analysis
- Sort Identified Gaps by Timing
- - Short-Term or Long-Term

Identified gaps need to be made explicit, and the intelligence collection effort must fill in the missing pieces of the model to answer the customer's questions.

PHASE2 ADVANTAGE © 2023 CHAPTER 05

Step 2: Develop the Collection Plan

- Evaluate the Available Resources
- Develop a Collection Strategy
- - Select the Appropriate Resources
- - Synchronize Collection to Requirements
- Develop *Specific Orders and Requests* (SOR)
- Prioritize the SORs for the Collection Assets

The goal is to develop a collection and tasking plan that produces the intelligence required to effectively answer the customer's information requirements.

PHASE2 ADVANTAGE © 2023 CHAPTER 05

Analyst and Collector Collaboration

- **Collectors**: Understanding of the Issues
- - Good Understanding = Effective Response
- **Analysts**: Understand Collection Capabilities
- - Analysts Lack the Knowledge of Collectors
- **Collectors**: Different Cultures and Focus
- **Analysts**: Understand Collector Preferences

The secret to any successful intelligence collection strategy is a sustainable and close relationship between analysts and collectors on the team.

PHASE2 ADVANTAGE © 2023 CHAPTER 05

Step 3: Task or Request the Collection

- Determine the Tasking Mechanism
- Determine the Request Mechanism
- Execute and Implement the Mechanisms
- - Begin the **Collection**
- - Begin the **Exploitation**

The benefit of communicating "collection intent" is the holistic perspective it provides internal and external collectors isolated from the planning process.

Step 4: Manage the Dissemination

- Arrange Direct Dissemination
- Determine Levels of Perishability
- Determine the Amount to Disseminate
- Identify the Media for Dissemination
- Disseminate the Collected Information

Whenever possible, write into the *Specific Orders and Requests* the requirement for direct dissemination of intelligence to the customer of the product.

Step 5: Evaluate the Reporting

- Monitor and Maintain the Synchronization
- Correlate Reports to the Requirements
- Screen the Reports for Relevance
- - Provide Feedback to the **Collectors**
- - Provide Feedback to the **Exploiters**

All *Specific Orders and Requests* must be fully satisfied in a timely manner, keeping the intelligence system fully synchronized.

Step 6: Update the Collection Planning

- Eliminate all Satisfied Requirements
- Redirect Assets to Unsatisfied Requirements
- Cue Assets to Collection Opportunities
- Maintain Continuous Synchronization
- Add New Requirements as Needed

Specific Orders and Requests are updated to keep intelligence synchronized with the operation, and collectors work only on unsatisfied requirements.

Overview of the Collection Strategy

- Gaps about the Target have been Identified
- The Nature of the Gap is Understood
- Not all Gaps are Collection Gaps
- - Data can be Inadequate or Nonexistent
- - Redundant Collection Wastes Time
- - Unneeded Efforts do not Solve Problems
- - Excess Data Compounds the Problem

When true collection gaps have been defined, it is time to plan for collection to fill them and offer guidance to the sources of collection.

Basic Collection Strategy Development

- Look at Successes from Previous Efforts
- - Collection Strategies against similar Targets
- - Measure Results of the Similar Strategies
- Cannot rely Solely on Previous Strategies
- - Predictable Collection can be Defeated
- - Innovative Approaches for Existing Assets
- Time the Collection for Desirable Content

Most collection strategies are developed as a result of past experiences; use of existing methods saves time, effort, and organizational resources.

Developing a Collection Plan

- Planning and Direction is the Foundation
- Customers of Intelligence Products Must:
- - **Articulate** Their Specific Needs
- - **Validate** Intelligence Requirements
- - **Address** Specific Questions
- - **Identify** Intelligence Gaps
- Collectors and Analysts cannot Guess Needs

"*Garbage in, garbage out*" is appropriate in understanding how effective the intelligence team is in responding to the customer's requirements.

PHASE2 ADVANTAGE © 2023 CHAPTER 05

The Collection Requirements Process

- **Organic Resources**
- - Intelligence Analysts
- - SOC Cybersecurity Personnel
- - Risk Management and Security
- **External Resources**
- - Upper Echelon Capabilities
- - Law Enforcement, Government, Academia

Intelligence *collection requirements* fall within two categories: tasking *organic resources* belonging to the organization or requesting use of *external resources*.

PHASE2 ADVANTAGE © 2023 CHAPTER 05

Chapter 05 Summary: Key Concepts

- The Role of the Collection Manager
- Collection Requirements Management Process
- Conducting a Gap Analysis
- Analyst and Collector Collaboration
- Overview of the Collection Strategy
- Developing a Collection Plan

PHASE2 ADVANTAGE © 2023 CHAPTER 05

Determining the Root Cause of Intelligence Issues

PHASE2 ADVANTAGE © 2023 CHAPTER 06

The Intelligence Issue

- The Issue must be Understood in Detail
- Determine the Need for the Request
- Allows for Decisions the Results will Support
- Success = An Accurate Issue Definition
- Discerns Important from Irrelevant Data

All intelligence analysis efforts begin with an *issue or problem definition* which aligns with the customer's interest about the target.

PHASE2 ADVANTAGE © 2023 CHAPTER 06

Veteran vs. Novice Analysts

- **Veterans**: Approach the Process Differently
- - Spend more Time Considering the Problem
- - Actively Avoid the **Framing Effect**
- - Use Knowledge of Previous Cases
- - Use Context for Creating Mental Models
- - Recognize when Information is Lacking
- - Focus on the Problem Definition Phase

The initial guidance customers give analysts about an issue is almost always incomplete and may be unintentionally misleading. Veteran analysts know this.

PHASE2 ADVANTAGE © 2023 CHAPTER 06

Preliminary Research Questions

- **1**: *Who is the **Customer**?*
- **2**: *What is the **Purpose**?*
- **3**: *What are the **Real** Questions?*
- **4**: *When are the **Answers** Required?*
- **5**: *What Product will be most **Effective**?*

In the discipline of scientific research, certain guidelines are generally followed before starting a new project which are based on five fundamental questions.

1: Who is the Customer?

- Identify the Intelligence Customers
- Understand their Specific Needs
- May Involve Several Intermediaries
- - Needs Tend to Become Distorted
- - Movement Through Long Channels
- Seek Direct Interaction with the Customer
- Consider Second-Order Recipients

After establishing parameters and obtaining guidance with these questions, analysts will have a starting point for defining the issue or problem statement.

2: What is the Purpose?

- Efforts Tend to have One Main Purpose
- The Purpose Should be Clear to Everyone
- - Clear when the Effort Begins
- - Clear in the Expected Result
- Customer Involvement Facilitates Clarity

On large and complex intelligence projects, the parameter statement will itself become a formal product to be disseminated to stakeholders.

3: What are the Real Questions?

- Obtain all Relevant Background Information
- - Underlying the Customer's Problem
- Understand the Problem's Real Impact
- - Impact on **Policy**
- - Impact on **Operational Decisions**
- Vaguely Worded Requests are Misleading
- **Result**: Not what Customer Wanted

The issue definition product helps to explain the real questions to be answered and all the related issues which need to be addressed.

PHASE2 ADVANTAGE © 2023 CHAPTER 06

4: When are the Answers Required?

- When the Product must be Delivered
- Many Products are Delivered too Late
- - Reports Received after Decisions are Made
- - Customer is Isolated from the Process
- The Target-Centric Model Speeds the Process
- - The Customer is Integrated into the Process

Once the issue definition product is complete, analysts will be able to focus on answering the questions that are relevant and meet customer expectations.

PHASE2 ADVANTAGE © 2023 CHAPTER 06

5: What Product will be most Effective?

- Written Reports are the Standard Format
- - Written (Electronic) Reports Endure
- - Easily Distributed to Multiple Customers
- Verbal Briefings are Another Option
- - Allows Customer Interaction and Feedback
- - Intended Recipient Receives the Message

A well-defined statement of the problem (the result of these answers) still requires two additional steps: *issue definition* and *issue decomposition*.

PHASE2 ADVANTAGE © 2023 CHAPTER 06

The Issue Definition

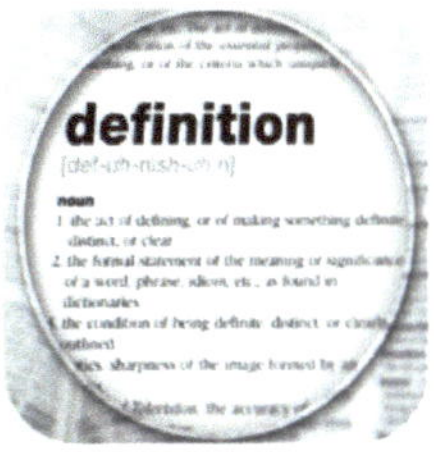

- Improperly Defined Issues Void Analysis
- Customers Rarely Ask the Right Questions
- - Not what They **Need** to Know
- - What They **Want** to Know
- - What They **Think** They Need to Know
- Analysts must Understand the Difference
- - Also Understand Gaps in Knowledge

When analysts receive the initial tasking (from a customer of self-initiated) time pressures may tempt them to give the *issue definition* minimal consideration.

PHASE2 ADVANTAGE © 2023 CHAPTER 06

Identifying Issue Categories

- **1**: Plans, Intentions, and Strategies
- **2**: Organizations and Leadership
- **3**: Operations
- **4**: Infrastructure
- **5**: Systems Acquisition and Implementation

Most intelligence needs, reflected in the questions customers will likely ask about a target, fall into one of five general *issue categories*.

PHASE2 ADVANTAGE © 2023 CHAPTER 06

#1: Plans, Intentions, and Strategies

- What is the Target's **Motivation**?
- What is the Target's **Doctrine** (Beliefs)?
- What Shapes the Target's **Perceptions**?
- What are the Target's **Priorities**?
- What are the Target's **Strategies**?
- What is Driving the Target's **Intent**?

Although the customer's initial question will likely be framed in terms of one of the categories, consider if it needs to be broadened to include other categories.

PHASE2 ADVANTAGE © 2023 CHAPTER 06

#2: Organization and Leadership

- Organizational Structure
- Identities of Key Actors
- Roles of Key Actors
- Purpose of Key Actors
- Internal Influence within the Organization
- External Influence outside the Organization

When a customer poses an intelligence question, there are often circumstances and conditions surrounding it the customer has not considered.

PHASE2 ADVANTAGE © 2023 CHAPTER 06

#3: Operations

- Individual and Organizational Activities
- Methods and Implementations
- Tools and Tactics
- Spatial Components
- Temporal Components

As intelligence analysts gain more experience, they become increasingly aware that customers at times do not know exactly what they are asking for.

PHASE2 ADVANTAGE © 2023 CHAPTER 06

#4: Infrastructure

- Utilities
- Bandwidth
- Computational Power
- Data Centers and/or Servers
- Transportation Networks
- Command and Control Centers
- Safe Houses

There are many times when a customer knows that an issue has potentially important implications but is unsure of the form the issue many take.

PHASE2 ADVANTAGE © 2023 CHAPTER 06

#5: Systems Acquisition and Implementation

- Research
- Development
- Training
- Capabilities
- Readiness
- Technical Parameters
- Vulnerabilities and Weaknesses

Having insight into a customer's priorities makes it easier to establish valid assumptions regarding their requirements for intelligence issues.

PHASE2 ADVANTAGE © 2023 CHAPTER 06

PMESII: Operational Values

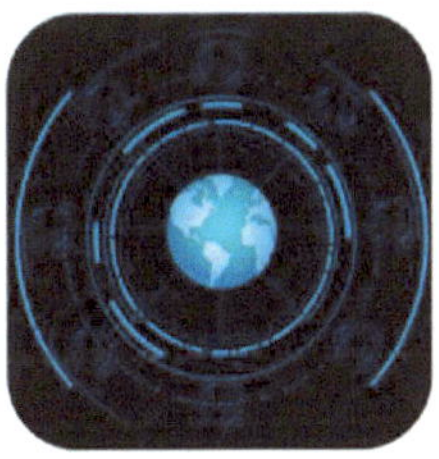

- **P**olitical (Responsibility and Power)
- **M**ilitary (Offensive Capability)
- **E**conomic (Individual and Group Actors)
- **S**ocial (Values, Behaviors, and Beliefs)
- **I**nfrastructure (Services Needed to Function)
- **I**nformation (Collect, Process, and Act)

Customers tend to think about actions. Intelligence analysts must think about both the adversary's actions **and** the likely impact of the customer's reactions.

PHASE2 ADVANTAGE © 2023 CHAPTER 06

ASCOPE: Environmental Factors

- **A**reas
- **S**tructures
- **C**apabilities
- **O**rganizations
- **P**eople
- **E**vents

The *ASCOPE* categories are environmental factors used to analyze the cultural and social conditions in which the operational variables (*PMESII*) exist.

PHASE2 ADVANTAGE © 2023 CHAPTER 06

Overview: ASCOPE / PMESII Matrix

OPERATIONAL VALUES

ENVIRONMENTAL FACTORS

	P Political	M Military	E Economic	S Social	I Information	I Infrastructure
A Areas	Areas Political	Areas Military	Areas Economic	Areas Social	Areas Information	Areas Infrastructure
S Structures	Structures Political	Structures Military	Structures Economic	Structures Social	Structures Information	Structures Infrastructure
C Capabilities	Capabilities Political	Capabilities Military	Capabilities Economic	Capabilities Social	Capabilities Information	Capabilities Infrastructure
O Operations	Operations Political	Operations Military	Operations Economic	Operations Social	Operations Information	Operations Infrastructure
P People	People Political	People Military	People Economic	People Social	People Information	People Infrastructure
E Events	Events Political	Events Military	Events Economic	Events Social	Events Information	Events Infrastructure

The *ASCOPE PMESII Matrix* allows analysts to identify known data, intelligence gaps, and focus on specific areas without losing the broader scenario view.

Example: ASCOPE / PMESII Matrix for **APT33**

OPERATIONAL VALUES

ENVIRONMENTAL FACTORS

	P Political	M Military	E Economic	S Social	I Information	I Infrastructure
A Areas	Regional Boundaries	Offensive, Defensive, and Current Operations		Dark Web, BBS, and Chat	Social Media and the Dark Web	Internet Service Providers
S Structures			Crypto Currency Exchanges			
C Capabilities	Local and National Political Influence	Strengths, Weaknesses, Opportunities, and Threats	Access to Funding Sources	Direct and Local Global (WEB)	Education, Training, and Technology	Ability to Maintain and Service Technology
O Organizations	Criminal, Legal, Political	Governmental and Nonstate	State Sponsored and Criminal	The Black Hat Consortium	TV Media and News Groups	Local Tech Companies
P People	Government and Provincial Leaders	Local, Proxy, National, and International				Programmers, Developers, Hardware, and HUMINT
E Events	Rallies, Protests, and Meetings	Exploit Events and Operations			Propaganda Recruiting Fundraising Campaigns	

This method considers the perspective of people, culture, and society in an intelligence effort to provide a holistic view of the tactical operating environment.

PHASE2 ADVANTAGE © 2023 CHAPTER 06

The Key Assumptions Check (I)

- **Risk**: Assumptions are not Questioned
- - Not Doubted at the Start of the Effort
- - Become Accepted as Fact Going Forward
- Identify any Key Assumptions
- Question the Validity and Relevance
- State Accepted Assumptions Explicitly
- Conduct a Relevancy Check when Finished

A *key assumption* is a hypothesis that has been accepted as true and will be a part of the problem definition or the final intelligence assessment product.

PHASE2 ADVANTAGE © 2023 CHAPTER 06

The Key Assumptions Check (II)

- *How much Confidence in the Assumption?*
- *How is the Confidence Level Justified?*
- *What can Undermine the Assumption?*
- *Could a True Assumption now be False?*
- *Are there Penalties for a Wrong Assumption?*
- *Is the Key Assumption a Key Factor?*
- *Has the Process Revealed New Assumptions?*

Customers can sometimes choose not to be informed by intelligence analysis on a given issue or to ignore challenges to their assumptions and/or opinions.

PHASE2 © 2023 CHAPTER 06

Issue Decomposition

- Presented as a Conceptual Framework
- Defines the Problem and/or Issue in Detail
- **Aka**: *Strategies-to-Task*
- **Aka**: *Problem Decomposition*
- **Aka**: *Decomposition and Visualization*

Issue decomposition is the technique for taking a seemingly intractable problem and breaking is down into a series of manageable sub-problems.

PHASE2 © 2023 CHAPTER 06

Issue Decomposition Taxonomy

- Start from the Issue Definition Statement
- Provide Specific Details about the Problem
- The Process Defines the Intelligence Needs
- Arrange from Top Level to Specific Task Level
- - Subtasks: **Key Intelligence Questions**
- The Approach has an Intuitive Appeal
- We Tend to Think about Issues Hierarchically

A *taxonomy* is a classification system in which objects are arranged into natural or related groups based on a factor common to each object in the group.

PHASE2 © 2023 CHAPTER 06

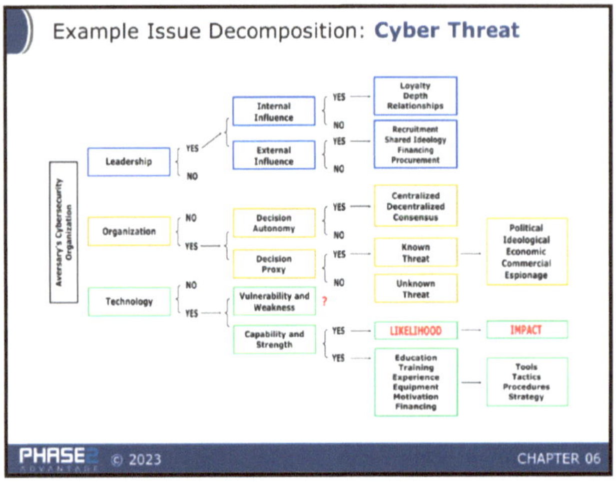
Example Issue Decomposition: Cyber Threat
Adversary's Cybersecurity Organization
Leadership
Organization
Technology
YES
NO
Internal Influence
External Influence
Decision Autonomy
Decision Proxy
Vulnerability and Weakness
Capability and Strength
Loyalty Depth Relationships
Recruitment Shared Ideology Financing Procurement
Centralized Decentralized Consensus
Known Threat
Unknown Threat
LIKELIHOOD
Education Training Experience Equipment Motivation Financing
Political Ideological Economic Commercial Espionage
IMPACT
Tools Tactics Procedures Strategy
PHASE2 ADVANTAGE © 2023
CHAPTER 06

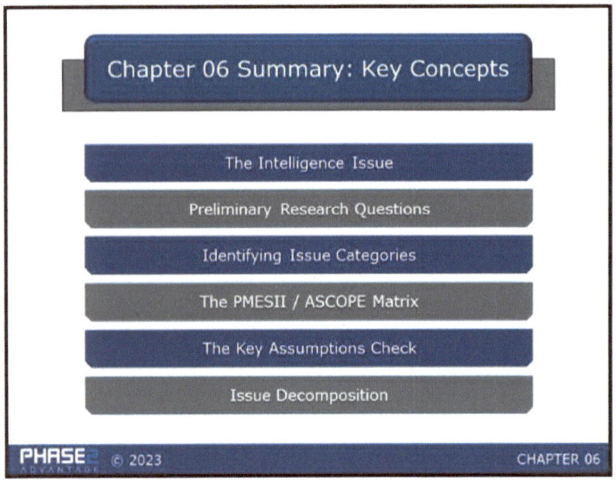
Chapter 06 Summary: Key Concepts
The Intelligence Issue
Preliminary Research Questions
Identifying Issue Categories
The PMESII / ASCOPE Matrix
The Key Assumptions Check
Issue Decomposition
PHASE2 ADVANTAGE © 2023
CHAPTER 06

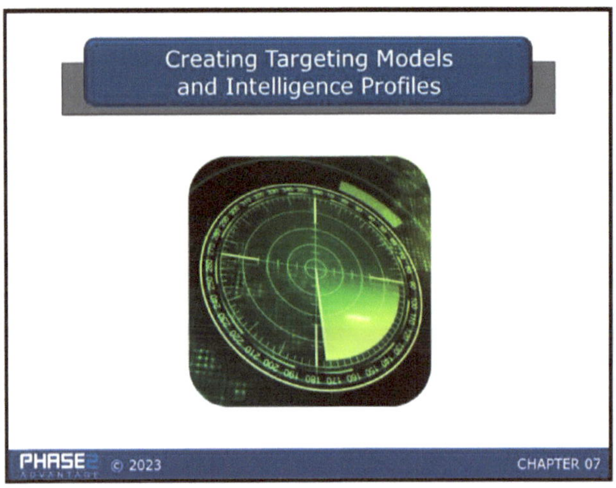
Creating Targeting Models and Intelligence Profiles
PHASE2 ADVANTAGE © 2023
CHAPTER 07

Modeling the Intelligence Target

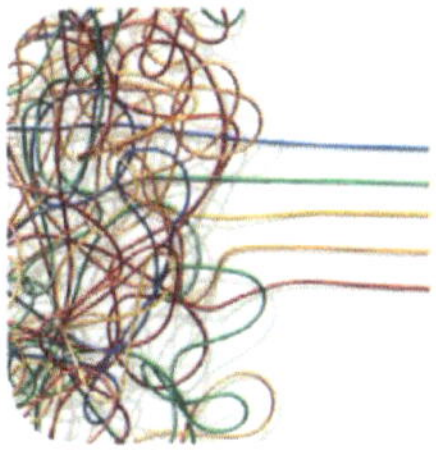

- A Model is a Replica or Representation
- - An Idea, Object, or System
- All Models Begin in the Human Mind
- Essential when Dealing with Complex Targets
- A Model is an Organizing Constraint
- - Facts, Hypotheses, and Assumptions
- A Form Useful for Making Decisions

Developing an intelligence assessment leads to the creation of a model (or set of models) of the target if the models do not already exist.

Intelligence Modeling Concepts

- **Physical Models**
- - A Tangible Representation of a Thing
- **Conceptual Models**
- - Creations of the Mind for Abstraction
- **Descriptive** (Describing the Representation)
- **Normative** (Prescribing a Course of Action)
- **Decision-Support** (Compare among ACHs)

Instead of interacting with the real system, an analyst can create a model which corresponds in specific ways to the actual target.

The Model Hierarchy

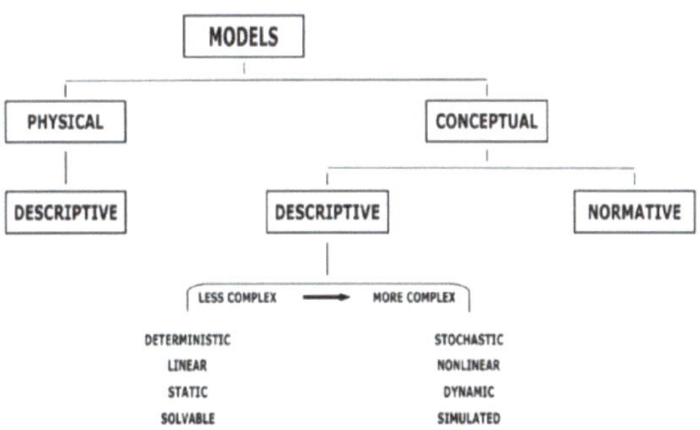

Visualization involves transforming an intelligence product into graphic, pictorial, or multimedia forms so the brain can process and understand complex data sets.

Descriptive Models (I)

- **Deterministic Models**
- - Relationships are Known and Specified
- **Stochastic Models**
- - Contains Uncertainties and Probabilities
- **Linear Models**
- - Linear Equations to Describe Relationships
- **Nonlinear Models**
- - Uses all Types of Mathematical Functions

Modeling does not always require a computer, and many useful models that analysts can utilize to solve complex issues exist only on paper.

PHASE2 ADVANTAGE © 2023 CHAPTER 07

Descriptive Models (II)

- **Static Models**
- - Ignore Time-Based Variances (Snapshots)
- **Dynamic Models**
- - Includes Spatial and Temporal Aspects
- **Solvable Models**
- - Analysis can be Used to Find Answers
- **Simulated Models**
- - "If Then" Scenarios to Predict Effects

Without a device to capture the full range of thinking and creativity that occurs in target modeling, analysts would have to keep track of too many details.

PHASE2 ADVANTAGE © 2023 CHAPTER 07

Comparative Models

- *Benchmarking*: Compares "Us to Them"
- **Parallel Lists**
- - Used for Comparing Alternative Solutions
- - Used for Comparative Intelligence Analysis
- **Matrices**
- - An Effective Way of Presenting Results
- - Can be Quantitative or Qualitative

Comparative techniques are a simple but useful form of modeling available to analysts that do not require sophisticated simulation exercises.

PHASE2 ADVANTAGE © 2023 CHAPTER 07

Example: APT Comparative Threat Matrix

Threat Matrix: APT Comparative Analysis					
Risk Values 1-20	**Known TTPs** Documented	**APT 28** Fancy Bear	**APT 32** OceanLotus	**APT 33** Elfin Team	**APT 41** Wicked Panda
1-20	Targets Petrochemical and Energy Industries	0	0	20	1
1-20	Targets Regions of Organizational Operation	0	0	20	1
1-20	Capable of Executing Zero-Day Exploits	2	4	5	6
1-20	Uses Boot or Auto-Start Execution	15	15	10	10
1-20	Targets Non-Standard Ports and/or Protocols	15	15	10	1
1-20	Uses Encryption for C2 and Data Exfiltration	15	15	15	10
1-20	Targets Windows Servers and File Sharing Systems	15	15	15	10
1-20	Uses Spear Phishing to Compromise Accounts	20	20	20	7
Total Value 1-160		82	84	115	46

Small Number Documented	Moderate Number Documented	Significant Number Documented
Risk Values		
1-4 / **Low Probability**	5-10 / **Moderate Probability**	11-20 / **High Probability**

Comparative Modeling: Disadvantages

- **Mirror Imaging**: Cultural Bias
- - Leads to Inaccurate Estimates and Analysis
- Models must be Specific to the Culture
- - Overlook Strengths and Weaknesses
- - Overlook Dependencies in the Network
- Create **Parallel Models** for Analysis
- - Assists in Analytical Benchmarking

One disadvantage of comparative modeling is that analysts may be inclined to rely on familiar models (*cultural bias*) instead of those of the target's culture.

Creating Intelligence Profiles

- Models of Individuals or Groups
- A Pattern of Mental and Behavioral Traits
- - Referred to as the **Modal Personality**
- Backgrounds can Create Specific Portraits
- - Observed Behavior and Biographical Data
- Can Predict the Behavior of Cyber Criminals
- - *Modal Personalities* Exist Equally Online

The purpose of creating *intelligence profiles* is to help predict what the profiled person or group will most likely do in various circumstances and conditions.

Pattern Models

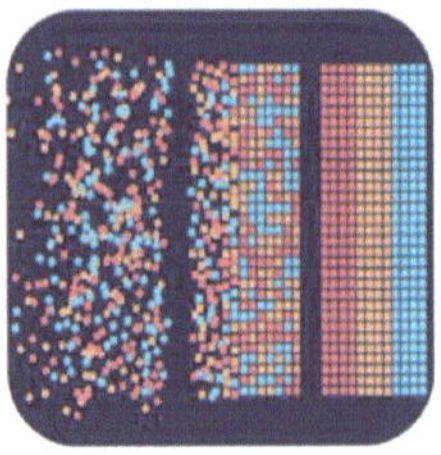

- Used to Recognize Patterns of Activities
- - A Departure or Change from the Known
- Software can Identify Discrete Trends
- Pattern Models Rely Heavily on Statistics
- - **Pattern-of-Life** (POL) Modeling
- - Useful in Identifying Future Activity
- Can be Done with Nonvisual Surveillance
- - Browsing History, Calls, and Transactions

Pattern modeling is a critical element of all intelligence as leaders, terrorists, and criminals have a modus operandi (a standard operational pattern).

PHASE2 ADVANTAGE © 2023 CHAPTER 07

Creating a Target Framework

- Define the System and Issues of Interest
- Select a Generic Target Framework
- **Structural System Models**
- - Actors, Objects, and Relationships
- **Functional System Models**
- - Concentrate on Achieved Results
- **Process System Models**
- - Focus on Interactions and their Dynamics

The generic target framework is the starting point for any analysis, as defining and deconstructing the target framework is the most important step in analysis.

PHASE2 ADVANTAGE © 2023 CHAPTER 07

Target Frameworks: Deconstructing Elements

- Creates a Visual Picture of the Target
- Commonly a Set of Interrelated Models
- Target Deconstruction Serves Many Purposes
- - Supports Analyzing Intelligence
- - Identify and Close Gaps in Knowledge
- - Develop a Finished Intelligence Product
- - What is Known, Unknown, and Assumed

Starting from the top-level view of the target, the analyst will *deconstruct* it into sub-categories (*elements*) and then sub-elements (all represented by models).

PHASE2 ADVANTAGE © 2023 CHAPTER 07

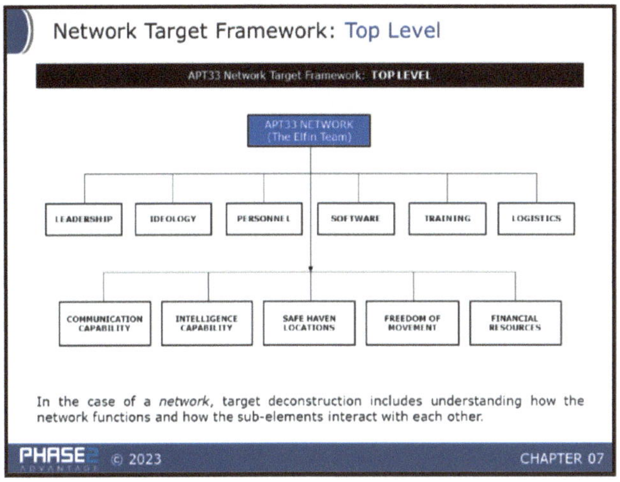

Network Target Framework: Top Level
APT33 Network Target Framework: TOP LEVEL
APT33 NETWORK (The Elfin Team)
LEADERSHIP
IDEOLOGY
PERSONNEL
SOFTWARE
TRAINING
LOGISTICS
COMMUNICATION CAPABILITY
INTELLIGENCE CAPABILITY
SAFE HAVEN LOCATIONS
FREEDOM OF MOVEMENT
FINANCIAL RESOURCES
In the case of a network, target deconstruction includes understanding how the network functions and how the sub-elements interact with each other.
PHASE2 ADVANTAGE © 2023
CHAPTER 07

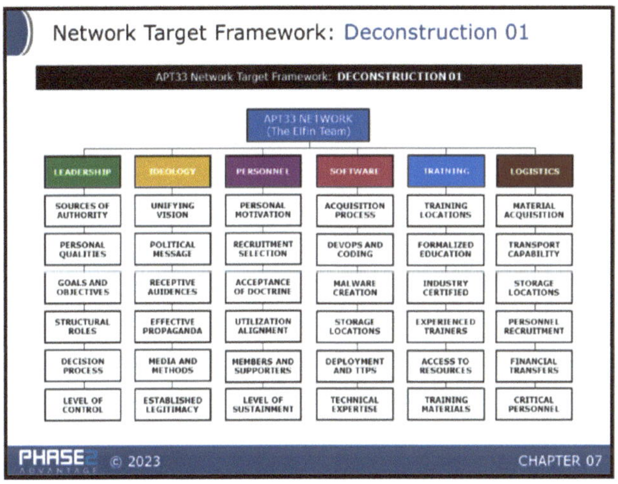

Network Target Framework: Deconstruction 01
APT33 Network Target Framework: DECONSTRUCTION 01
APT33 NETWORK (The Elfin Team)
LEADERSHIP
SOURCES OF AUTHORITY
PERSONAL QUALITIES
GOALS AND OBJECTIVES
STRUCTURAL ROLES
DECISION PROCESS
LEVEL OF CONTROL
IDEOLOGY
UNIFYING VISION
POLITICAL MESSAGE
RECEPTIVE AUDIENCES
EFFECTIVE PROPAGANDA
MEDIA AND METHODS
ESTABLISHED LEGITIMACY
PERSONNEL
PERSONAL MOTIVATION
RECRUITMENT SELECTION
ACCEPTANCE OF DOCTRINE
UTILIZATION ALIGNMENT
MEMBERS AND SUPPORTERS
LEVEL OF SUSTAINMENT
SOFTWARE
ACQUISITION PROCESS
DEVOPS AND CODING
MALWARE CREATION
STORAGE LOCATIONS
DEPLOYMENT AND TTPS
TECHNICAL EXPERTISE
TRAINING
TRAINING LOCATIONS
FORMALIZED EDUCATION
INDUSTRY CERTIFIED
EXPERIENCED TRAINERS
ACCESS TO RESOURCES
TRAINING MATERIALS
LOGISTICS
MATERIAL ACQUISITION
TRANSPORT CAPABILITY
STORAGE LOCATIONS
PERSONNEL RECRUITMENT
FINANCIAL TRANSFERS
CRITICAL PERSONNEL
PHASE2 ADVANTAGE © 2023
CHAPTER 07

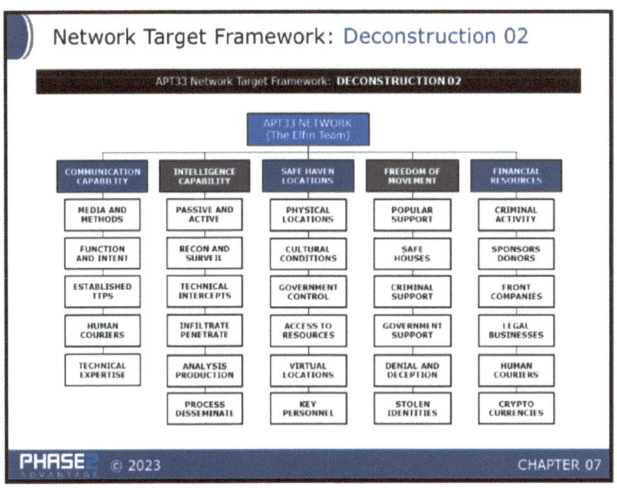

Network Target Framework: Deconstruction 02
APT33 Network Target Framework: DECONSTRUCTION 02
APT33 NETWORK (The Elfin Team)
COMMUNICATION CAPABILITY
MEDIA AND METHODS
FUNCTION AND INTENT
ESTABLISHED TTPS
HUMAN COURIERS
TECHNICAL EXPERTISE
INTELLIGENCE CAPABILITY
PASSIVE AND ACTIVE
RECON AND SURVEIL
TECHNICAL INTERCEPTS
INFILTRATE PENETRATE
ANALYSIS PRODUCTION
PROCESS DISSEMINATE
SAFE HAVEN LOCATIONS
PHYSICAL LOCATIONS
CULTURAL CONDITIONS
GOVERNMENT CONTROL
ACCESS TO RESOURCES
VIRTUAL LOCATIONS
KEY PERSONNEL
FREEDOM OF MOVEMENT
POPULAR SUPPORT
SAFE HOUSES
CRIMINAL SUPPORT
GOVERNMENT SUPPORT
DENIAL AND DECEPTION
STOLEN IDENTITIES
FINANCIAL RESOURCES
CRIMINAL ACTIVITY
SPONSORS DONORS
FRONT COMPANIES
LEGAL BUSINESSES
HUMAN COURIERS
CRYPTO CURRENCIES
PHASE2 ADVANTAGE © 2023
CHAPTER 07

Alternative Frameworks

- Keep Multiple Target Frameworks in Mind
- Useful for Conflicting Intelligence Information
- Multiple Alternative Models are Needed
- Analytic Pitfalls can Undermine Assessments
- - Failure to Consider Alternative Scenarios
- - **Ego**: Rejection of Dissenting Views
- - **Time**: Tight Deadlines Force Easy Choices
- - **Customers**: Choose the most Favorable

Alternative frameworks and their supporting models are valuable assets for analysts to use and an essential part of the intelligence analysis process.

Competitive Frameworks

- Analysts are Dealing with Uncertainty
- Different Analysts have Different Conclusions
- Customers may Provide Competing Analysis
- Be Inclusive when Defining Frameworks
- Alternatives not Offered may be Correct

It is an established standard in intelligence that, if the organization can afford the resources, independent groups should provide competing analysis.

Chapter 07 Summary: Key Concepts

- Modeling the Intelligence Target
- Intelligence Modeling Concepts
- Creating Intelligence Profiles
- The Model Hierarchy
- Creating a Target Framework
- Target Frameworks: Deconstructing Elements

Evaluating Intelligence Sources and Evidence

Issue Decomposition and the Target Framework

- **Decomposition**: Ask the Right Questions
- **Target Model**: Provides the Right Context
- Similar but Separate Models
- - They Come from Different Sources
- - They Change in Different Ways
- - Issue may Address a Portion of the Model
- - A Model may Serve Several Issues

The *issue decomposition* and the *target framework* are separate concepts and separate models, but they are closely related to each other.

Analyzing Existing Intelligence

- Evaluate the Intelligence for **Relevance**
- Evaluate the Intelligence for **Credibility**
- *Collate* the Intelligence into the Target Model
- - Review Existing Finished Intelligence
- - Examine the Existing Raw Intelligence
- - Acquire New Raw Intelligence
- - Evaluate the New Raw Intelligence
- - Combine in the Generic Target Framework

In intelligence analysis, *collation* is the organizing of relevant information in a coherent way, taking both the source and the context into consideration.

Existing Finished Intelligence

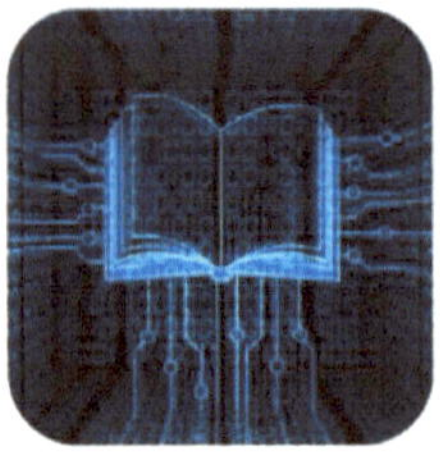

- Start with Existing Reports on File
- - Most Issues have been Addressed
- The First Step is a Literature Search
- Establish the Current State of Knowledge
- Identify Controversies and Disagreements
- Do not Accept Existing Intelligence as Fact
- Do a *Key Assumptions Check* on the Model
- Review Existing Raw Intelligence Data

Prior to starting an intelligence collection effort, analysts should ensure that they are already aware of information already collected on a target.

Sources of Raw Intelligence

- Intelligence has Two Primary Source Types
- **Literal** and **Nonliteral**
- *Literal*: In a Form Used for Communication
- - Understood by Analysts and Customers
- - Material Requires no Special Exploitation
- *Nonliteral*: Requires Special Processing
- Not a Perfectly Clean Distinction
- Treated Differently Depending on the Source

Successful analysis requires an understanding of the sources of intelligence, and the nature and limitations available from those sources.

Evaluating Evidence

- The Process Involves Subjective Judgments
- Conduct a **Quality of Information Check**
- **1**: Evaluate the *Source*
- **2**: Evaluate the *Channel*
- **3**: Evaluate the *Evidence*
- Use the Concept of **Occam's Razor**
- - Explain with the Fewest Hypotheses
- **Risk**: Answer may be Highly Complex

The fundamental objective in weighing intelligence evidence is determining its credibility; the analyst must evaluate it for completeness and veracity.

Evaluating the Source

- Accept Nothing at Face Value
- Evaluate the Origin of Evidence Carefully
- Be Aware of Potential Motives
- *- Is the Source Competent?*
- *- Does the Source have Required Access?*
- *- Does the Source have a Vested Interest?*

It is possible for analysts to make information fit almost any desired conclusion, especially by selectively disregarding conflicting or inconvenient facts.

 CHAPTER 08

Evaluating the Communication Channel (I)

- The Message Gets Distorted in Transit
- Randomness Increases in any System
- - Accuracy Decreases as Length Increases
- - Accuracy Decreases as Nodes Increase
- **Primary Sources**: Firsthand Information
- **Secondary Sources**: Intermediaries
- Entropy has an Impact on Intelligence
- *- "The Creeping Validity Problem"*

Over a long chain of human communication (especially in bureaucracies), the received message may bear little resemblance to the information originally sent.

 CHAPTER 08

Evaluating the Communication Channel (II)

- **Ask Questions about the Channel**:
- *- What was the Channel?*
- *- Was Information Provided Intentionally?*
- *- What Part of the Information is True?*
- *- Could it be Deception and Denial?*
- *- Could it be a Signal to the Adversary?*
- *- If it is a Signal, what is the Reason?*

Instead of a "primary-versus-secondary" distinction, it is more important for the intelligence analyst to consider the communication channel itself.

 CHAPTER 08

Source Analysis

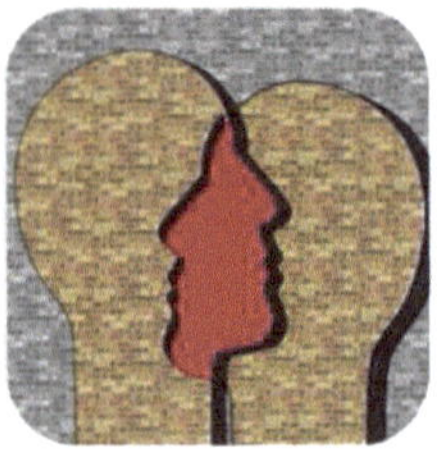

- Consider the Past Reporting History
- Consider the Psychological Profile
- An Incomplete Picture is Misleading
- Counterarguments Tend to be Avoided
- Analysts Take Steps to Find Missing Evidence
- Missing Evidence can Complete the Picture

The synergistic effect of combining information from multiple sources serves to strengthen conclusions and increase the team's confidence in them.

PHASE2 ADVANTAGE © 2023 CHAPTER 08

Source Reliability

- **A**: *Reliable* (History of Complete Reliability)
- **B**: *Usually Reliable* (Minor Doubts)
- **C**: *Fairly Reliable* (Doubts)
- **D**: *Not Usually Reliable* (Significant Doubts)
- **E**: *Unreliable* (Lacks Authenticity or Invalid)
- **F**: *Unknown* (Insufficient Information)

The *source reliability* is rated between "A" (complete reliability) to "E" (invalid information), with "F" for a source without a sufficient history to rate accurately.

PHASE2 ADVANTAGE © 2023 CHAPTER 08

Information Reliability

- **1**: *Confirmed* (by Independent Sources)
- **2**: *Probably True* (Logical, not Confirmed)
- **3**: *Possibly True* (Logical, not Confirmed)
- **4**: *Doubtful* (not Logical, not Confirmed)
- **5**: *Improbable* (not Logical and Contradicted)
- **6**: *Unknown* (Validity cannot be Determined)

The *information reliability* is rated between "1" (confirmed) to "5" (improbable), with a "6" for information that cannot be confidently evaluated by the analyst.

PHASE2 ADVANTAGE © 2023 CHAPTER 08

Vested Interest and Bias

- Government Agencies have Similar Problems
- Officials may Distort or Conceal Facts
- - To Support Policy Decisions
- - To Protect Personal Interests
- May Exaggerate Situational Importance
- - Encourage more Government Funding
- Tests to Prove or Disprove may be Slanted

Analysts occasionally encounter the "professional source" who sells information to the highest bidder and makes it as interesting as possible.

Evaluating the Credentials of Evidence

- **Fact**
- - Verified Information Known to Exist
- **Direct Information**
- - Likely Factual as a Result of the Source
- **Indirect Information**
- - Veracity of Truth has not been Established
- **Risk**: "Direct Information" may Mislead
- - *Denial and Deception Operations*

Credibility refers to the extent to which an analyst can believe something, and *relevance* means the information has value in addressing an issue or question.

Evaluating Evidence: Disadvantages

- **1**: Vividness Weighting
- **2**: Weighing Based on the Source
- **3**: Favoring the Most Recent Evidence
- **4**: Favoring or Disfavoring the Unknown
- **5**: Trusting Hearsay
- **6**: Relying on Expert Opinions
- **7**: Premature Closure
- **8**: Philosophical Predisposition

Customers place priority on evidence they collect themselves; a mistake that leadership can make easily which causes them to be vulnerable to deception.

Combining Types of Evidence

- **Divergent Evidence**
- - Two Data Sets Favor Different Conclusions
- **Convergent Evidence**
- - Two Data Sets Favor the Same Conclusion
- **Corroborative Redundancy**
- - Known Information is Corroborated
- **Cumulative Redundancy**
- - Information is Validated and Enhanced

The intelligence analysis process involves combining disparate types of evidence and helps analysts to make qualitative judgments about conclusions to support.

Analysis of Competing Hypotheses (ACH)

- Can Assist Judgment on Important Issues
- Carefully Weigh Alternative Conclusions
- - Helps to Minimize Cognitive Limitations
- - Identifies a Complete Set of Hypotheses
- - Evaluates Consistent and Inconsistent Data
- Analyze Potential Analysis Gaps
- View the Target from Alternative Perspectives
- Assists in the Questioning of Assumptions

It is often important to combine evidence and demonstrate the logical process of reaching a conclusion based on that evidence through *structured argumentation*.

Combining Evidence with Bayesian Analysis

- New Information is Evaluated Individually
- - Combined with *Historical* Assessment
- - Combined with *Subjective* Assessment
- Determine the New Likelihood of Occurrence
- Observed Data are Attributable to Causes
- **Risk**: Difficult to Teach and Apply in Practice

Bayesian analysis uses new data to modify previously estimated probabilities and can therefore be used to narrow the error bounds on estimates.

The Role of Information Technology

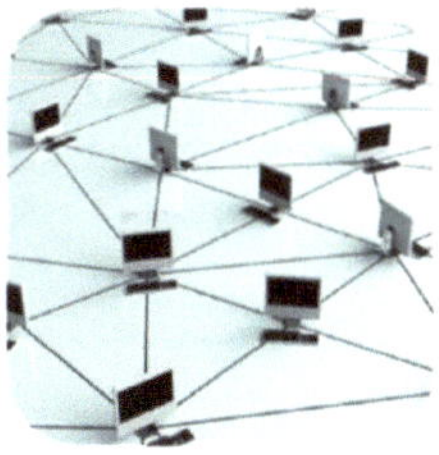

- A Challenge to Deal with Large Datasets
- A Challenge to Deal with Complex Datasets
- Database Management Tools are Ineffective
- Collected Data is too Large to be Analyzed
- Collected Data is too Large to be Stored
- - Palantir Technologies
- - IBM i2 Analyst's Notebook
- - Visual Link

A significant consequence of the information explosion is that analysts must now deal with "big data" when collating and analyzing intelligence products.

Chapter 08 Summary: Key Concepts

- Issue Decomposition and the Target Framework
- Analyzing Existing Intelligence
- Sources of Raw Intelligence
- Evaluating the Credentials of Evidence
- Combining Types of Evidence
- The Role of Information Technology

Predicting the Adversary's Planning and Operations

Anticipatory Analysis (I)

- Requires Another Conceptual Framework
- Referred to as **Probabilistic Analysis**
- Requires Highly Structured Thinking
- Presents a View of what is Likely to Happen
- Takes the Form of an Intelligence Estimate
- - Characterizes and Reduces Uncertainty
- - Reveals Several Potential Outcomes
- - Enhances Resiliency and Adaptability

Estimating the future state of a target can only be done using probabilities, so the risk of a mistake is greater than when analysts rely on established facts.

Anticipatory Analysis (II)

- Customers will Accept Anticipatory Analysis
- - Well-Reasoned and Understandable
- - Logical Development can be Understood
- Analysts must Apply a Formal Methodology
- Customer can See the Basis for Conclusions
- Identify Forces Likely to Affect the Target
- Referred to as "*Outside In Thinking*"
- - Critical External Factors are Identified

Most intelligence customers tend to focus on current situations, and those who take a strategic view are inclined to rely on their own predictive abilities.

Anticipatory Analysis (III)

- List all External Key Factors
- - Those that Little Influence can be Exerted
- List Key Factors that can be Influenced
- - Include the Behavior of Allies / Adversaries
- Assess all the External Key Factors
- - Assess the Affect on the Analytic Issue
- Determine if Factors have an Impact
- - Base Assessment on Available Evidence

The most important contribution at the start of an estimate is the assessment of factors that will shape future events and the future state of the target model.

Background Forces

- **Inertia**
- - The Tendency to Resist Change
- **Opposition**
- - Opposing or Resistive Forces
- **Contamination**
- - The Degradation of any External Factor
- **Feedback**
- - The System Adapts, Learns, and Changes

Some forces are both broad and subtle; they operate as a background process on most targets and need to be considered by the analyst.

Synergistic Considerations

- **Synergy**
- - The Whole is more than the Sum of Parts
- Synergy is **not** a Force
- - The Result of the Interaction of Forces
- A Result of Cooperative Efforts and Alliances
- Synergy has a "*Force Multiplier Effect*"

It is essential that the intelligence analyst develop an understanding of a broad range of concepts in order to function in a multidisciplinary environment.

Causal Models

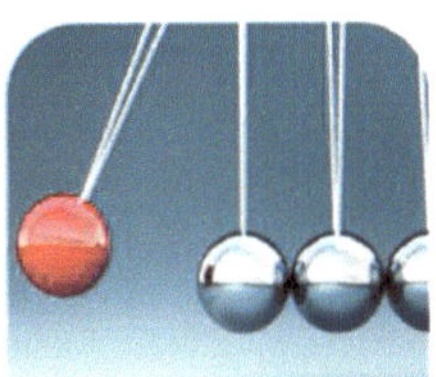

- They are Composed of Nodes
- The Nodes are Usually Variables
- The Relationship can be Negative or Positive
- They Use Probability Laws and Distributions
- - The Time Sequence must be Proper
- - There must be a Theory of Causality
- - The Relationship cannot be Questionable
- - The Analyst must Prove Covariation Exists

Causal models depict elements of the target, elements affecting the target, and describe how those elements are related by probabilities.

Anticipatory Analysis Methodology

- Uses Past and Current Target Models
- Helps to Assess the Potential Future State
- **1**: Identify Forces Acting on an Entity
- **2**: Associate Likely Future Forces
- **3**: Predict a Range of Likely Changes
- **4**: Consider Old and New Forces over Time
- **5**: Indicator of Confidence in Judgments

In addition to taking a multidisciplinary approach, the second critical trait of a successful predictive analyst is the ability to establish proper historical context.

The Estimative Approach

- **Extrapolation**
- - Assumes the Forces do not Change
- **Projection**
- - Assumes the Forces do Change
- **Forecasting**
- - Assumes that New Forces are Added

The methodology (conceptual framework) of anticipatory intelligence relies on three predictive mechanisms: *extrapolation*, *projection*, and *forecasting*.

The Estimative Methodology

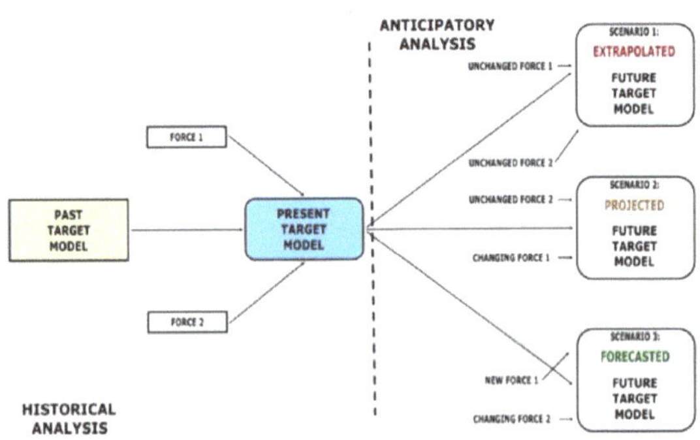

Extrapolation

- Little Uncertainty about the Current State
- Confidence in Forces Acting on the Target
- New Information is Assigned Low Priority
- It is Typically Accurate in the Short-Term
- - Assumes an Accurate Starting Point
- - Knowledge of Direction of Movement

An *extrapolation* is a statement, based only on past observations, of what is expected to happen. It is the most conservative method of prediction.

 CHAPTER 09

Projection

- Makes Use of Two Analytic Techniques
- **Force Analysis**
- - A Qualitative Study of the Forces
- **Probabilistic Reasoning**
- - A Systematic Method of Reasoning
- - Subjective Estimates of Probabilities
- - Makes Estimates Explicit and Consistent

Projection predicts a range of likely future outcomes based on the assumption that forces that have affected the target in the past will change.

 CHAPTER 09

Generating Alternative Outcomes

- Outcomes Defined as *Alternative Scenarios*
- Alternatives Should be Mutually Exclusive
- Alternatives Should be Exhaustive
- Listing Outcomes can Expand Perspective
- Generates Useful Insights into Issues
- List all Outcomes that are Relevant
- Helps to Prevent *Premature Closure*
- Adds Structure to Complicated Issues

The first step in probabilistic reasoning is formally stating that more than one outcome is possible (*alternatives*) by listing all foreseeable possible scenarios.

 CHAPTER 09

Influence Tree Diagrams

- Customer Requires an Estimate of Outcomes
- Three Possible Outcomes are Defined
- **1**: APT33 Successfully Compromises a Target
- **2**: Defense Against APT33 is Successful
- **3**: Status of Outcome Remains Unchanged
- Describe Forces Affecting Possible Outcomes
- *"Driver Events"* aka: **Linchpin Premises**
- Assumptions must be Explicit

Influence tree diagrams represent a systematic approach to force analysis and present a unique form of causal model for the intelligence analyst and customer.

Example: Influence Tree Diagram

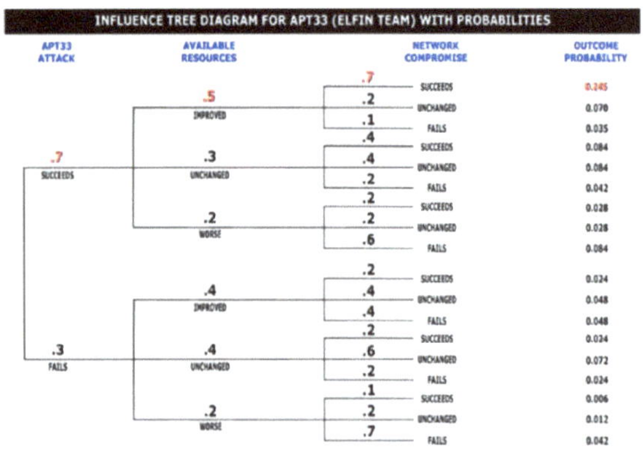

Influence Net Modeling Diagrams

- Combines Two Methods of Decision Analysis
- **1**: *Bayesian Inference Net Analysis*
- **2**: *Influence Diagramming Techniques*
- The Analyst Defines "Influence Nodes"
- - Arrowhead (Positive) and Circle (Negative)
- An Important Tool in Target-Centric Approach
- The Diagram can be Shared with Customers
- Encourages Customers to Provide Feedback

Influence net modeling is a powerful tool for projection of complex target models where the influence tree would be too cumbersome for practical analytical use.

Example: Influence Net Diagram

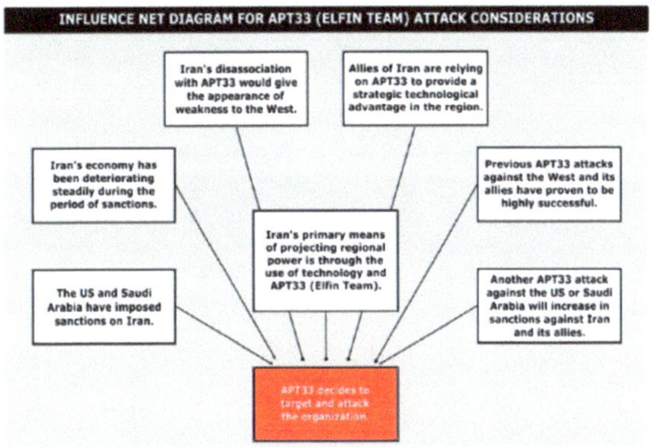

Forecasting

- Critical for Effective Strategic Decisions
- Attempts to Identify New Forces
- Considers Possible Effects of Developments
- Background Forces must be Considered
- - Inertia, Opposition, Synergy, and Feedback
- High-Level Anticipatory Analysis is Preferred

A primary objective of *forecasting* in intelligence operations is to define possible alternative outcomes of the target framework, not just the most likely ones.

Evaluating Forecasts

- **1**: Clarity
- **2**: Credibility
- **3**: Plausibility
- **4**: Relevance
- **5**: Urgency
- **6**: Comparative Advantage
- **7**: Technology Quality

Intelligence analysts should advise customers that forecasts are dynamic and need constant adjustment to be helpful in guiding thoughts and actions.

High-Impact / Low-Probability Analysis

- Used for Sensitizing Customers and Analysts
- Consequences of Unlikely Developments
- Typically Arise from Divergent Phenomena
- Unexpected and Unpleasant Surprises
- Describes how Development Might Begin
- Considers Plausibility and Consequences
- Referred to as a **Demonstration Scenario**

Projections and forecasts focus on the most likely outcomes, but customers also need to be aware of unlikely outcomes with severe adverse effects.

Chapter 09 Summary: Key Concepts

- Anticipatory Analysis
- Background Forces
- Anticipatory Analysis Methodology
- Extrapolation, Projection, and Forecasting
- Influence Tree and Influence Net Diagrams
- High-Impact / Low-Probability Analysis

Structured Techniques for Conducting Analysis

Structured Analytic Techniques (I)

- Helps Make Sense of Complex Problems
- Compare and Weigh Pieces of Information
- Ensures Analysts Focus on the Issue
- Forces Consideration of Only One Element
- Overcome Logical Fallacies and Biases
- Identification of Patterns and Correlations
- Enhances Data Gathering and Review
- Effective Thinking to Derive Alternatives

PHASE2 ADVANTAGE © 2023 CHAPTER 10

Structured Analytic Techniques (II)

- Sorting
- Matrices
- Threat Intentions Matrix
- Event Mapping
- Event Trees
- Subjective Probability

These techniques will improve analyses by making them more rigorous, provide ways to measure progress, and improve the presentation of the final product.

PHASE2 ADVANTAGE © 2023 CHAPTER 10

Sorting

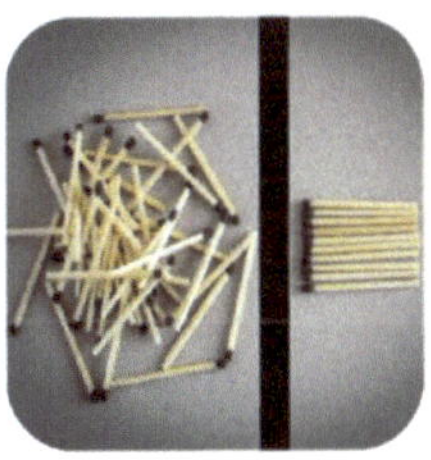

- Useful for Reviewing Massive Datasets
- Aids in the Review of Multiple Categories
- Can Present Possible Trends
- - Similarities, Differences, and Insights
- Effective During Initial Data Gathering
- Effective for Initial Hypothesis Generation
- Standardizing the Data is Imperative
- **Risk**: Improper Sorting can Hide Insights

Sorting is effective when information elements can be broken out into categories for comparison using automated computer software.

PHASE2 ADVANTAGE © 2023 CHAPTER 10

Sorting: Methodology

- Review the Information Categories
- Place the Data into a Spreadsheet
- List all the Facts Involved Used for Sorting
- Uncovers Possible Patterns and Groupings
- Groupings Illustrate the Schema Categories
- Group the Items According to Schema
- Choose a Category and Sort the Data
- Review and Re-Review the Sorted Data

Sorting massive amounts of data can provide insights into trends or anomalies that warrant further investigation that may have otherwise gone unnoticed.

Matrices

- Useful for Isolating Critical Data
- - Abundant Amounts of Relevant Information
- Enables Analytic Focus on Each Option
- Allows Issue Elements to be Separated
- Issues Categorized by Type for Comparison
- Helps to Identify Patterns and Correlations
- An Intermediate Step in **Link Analysis**

Matrices are useful whenever there are more options or more intricate data than can be conceptualized at one time without a visual representation.

Matrices: Methodology

- Draw a Matrix Using Columns and Rows
- Enter the Range of Data or Criteria
- Note the Relationships at Grid Intersections
- Review the Hypotheses Against the Matrix
- Develop New Hypotheses if Required
- Develop a Template for Recurring Topics
- Color-Code Results to Aid Understanding

The two-dimensional design of matrices limits their use for collating data on complex issues and leaving out pertinent data will oversimplify issues.

Matrices: Example

OPERATIONAL VALUES

ENVIRONMENTAL FACTORS

	P Political	M Military	E Economic	S Social	I Information	I Infrastructure
A Areas	Areas Political	Areas Military	Areas Economic	Areas Social	Areas Information	Areas Infrastructure
S Structures	Structures Political	Structures Military	Structures Economic	Structures Social	Structures Information	Structures Infrastructure
C Capabilities	Capabilities Political	Capabilities Military	Capabilities Economic	Capabilities Social	Capabilities Information	Capabilities Infrastructure
O Operations	Operations Political	Operations Military	Operations Economic	Operations Social	Operations Information	Operations Infrastructure
P People	People Political	People Military	People Economic	People Social	People Information	People Infrastructure
E Events	Events Political	Events Military	Events Economic	Events Social	Events Information	Events Infrastructure

The matrix above shows one method of cross-walking the operational variables memory aid **PMESII** and the environmental factors memory aid **ASCOPE**.

Threat Intentions Matrix

- Helps to Mitigate Analyst Biases
- Insight into Threat's Decision-Making Criteria
- Develops Clear Indicators for Each Option
- Allows for Specific Collection Planning
- All Alternatives and Options must be Entered
- Can be Completed Rather Quickly
- Better Options may Become Apparent
- The Criteria Used is not Extensive

The *threat intentions matrix* is an analytic technique used by intelligence analysts to efficiently look at information from the threat's point of view.

Threat Intentions Matrix: Methodology

- Enter Decision Options from Threat's View
- Enter the Objectives for Each Option
- Enter the Benefits of the Threat's Decisions
- Enter the Risk of the Threat's Options
- Complete the *Implications Column*
- - **Transition** from Threat to Analyst View
- Enter Implications from the Analyst's View
- Enter the Indicators from the Analyst's View

It is necessary for the analyst using this technique to have knowledge of the motivation, goals, and objectives of the threat making the decisions.

Example Threat Intentions Matrix: **APT33**

Options	Objectives	Benefit	Risk	Implications	Indication
Organization's Public-Facing Website	Remove the ability to interact with the general public	Show of power and an embarrassment to the company	Mission failure, public embarrassment, and risk of sanctions	APT33 gains stature, Iran gains influence, new global sanctions	Ping scans, ping sweeps, and signs of potential DDoS
Organization's ICS / SCADA Control System	Destroy the ability to conduct critical operations	Show of power, media attention, and aid recruitment	Adequate training and maintaining security during the attack	APT33 gains stature, new sanctions, energy shortage	Increased phishing, actions consistent with ATT&CK Matrix
Organization's IAM / Password Database	Compromise the integrity of the credentialing system	Long-term access to sensitive data and accounts	Adequate training and maintaining security during the attack	APT33 gains stature, disrupt security, impact operations	Signs of network enumeration and increased phishing
Organization's Payroll / Finance Database	Disrupt the ability to conduct vital financial transactions	Show of power, disrupt operations and employee morale	Adequate training and maintaining security during the attack	APT33 gains stature, disrupt operations, loss of employees	Signs of network enumeration and spear phishing
Organization's Human Resource Database	Compromise and/or destroy employee personnel data	Acquire data on key employees for targeting efforts	Adequate training and maintaining security during the attack	APT33 gains stature, disrupt security, impact operations	Signs of network enumeration and spear phishing
Organization's Client Services Database	Compromise and/or destroy client business data	Show of power, disrupt operations and client confidence	Adequate training, security during attack and client response	APT33 gains stature, new sanctions, local economy impacted	Signs of network enumeration and attacks on clients
Organization's User Endpoint Devices	Establish a foothold for a large-scale network compromise	Acquire data about users and daily operations	Mission failure, public embarrassment, and impact recruitment	APT33 gains stature, disrupt security, impact operations	Increased phishing, event-based alerting, and help desk calls

It is important to enter the information for every decision-making criteria in a column before moving to the next in order to mitigate bias.

Event Mapping

- Encourages a Brainstorming Approach
- Promotes Creativity in Generating New Ideas
- See Associations not Previously Discovered
- Elements Arranged Intuitively by Importance
- Organized into Groups, Branches, and Areas
- Aids in the Recall of Existing Memories
- **Risk**: Can Become Overly Detailed
- **Risk**: Including Irrelevant Events and Ideas

Event mapping uses a brainstorming diagram to represent the scenarios in the hypotheses linked around a central phrase representing the issue to be analyzed.

Event Mapping: Methodology

- Use a Blank Sheet of Paper on a Whiteboard
- Key Words or Phrases that Represent Ideas
- List Ideas as They Occur, Wherever They Fit
- Do not Prejudge Ideas or Use Restraint
- Develop the Directions for the Topics
- Increase Detail as the Map Expands
- Use Arrows to Show Links Between Events
- Provide Context by Writing on the Links

Use this technique when a nonlinear method is desired to generate, visualize, structure, and delineate the events in a scenario or related issue hypotheses.

Example Event Mapping: **APT33 Attack**

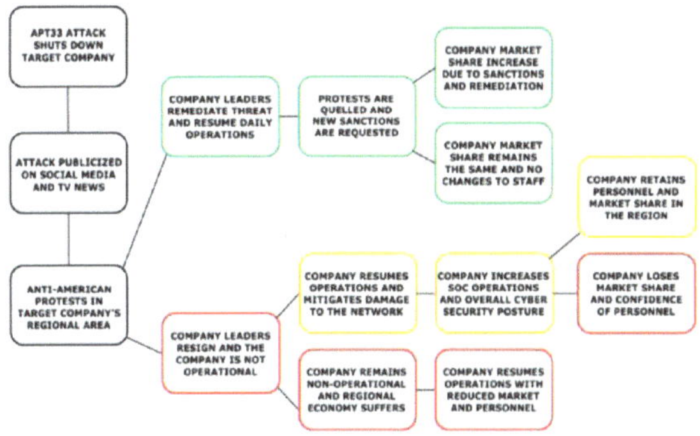

PHASE2 ADVANTAGE © 2023 CHAPTER 10

Event Trees

- Used to Clarify Alternative Event Sequences
- List Multiple and Mutually Exclusive Options
- Cover the Spectrum of Alternatives
- Depicts Threat Options with Decision Points
- Gives Insight into Potential Vulnerabilities
- Clarifies the Presumed Event Sequences
- Decisions Between Initial Event and Outcome
- Helps to Determine Collection Requirements

An *event tree* is a graphic depiction of a possible sequence of events, including junctures within the event sequence, depicting threat options and decisions.

PHASE2 ADVANTAGE © 2023 CHAPTER 10

Example Event Tree: **APT33 Attack**

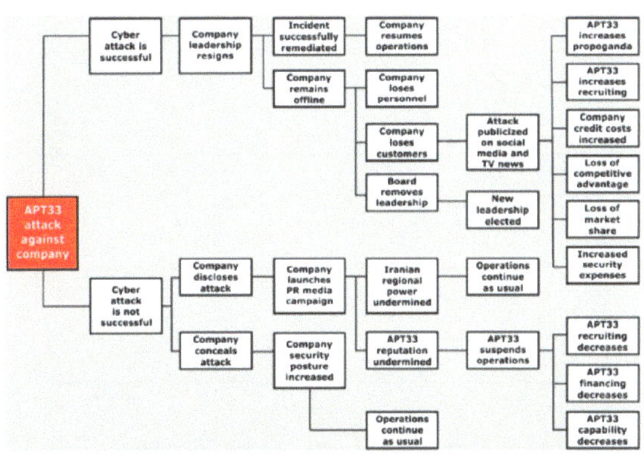

PHASE2 ADVANTAGE © 2023 CHAPTER 10

Subjective Probability

- Quantifiably Expresses a Degree of Belief
- Compares the Likelihood of Hypotheses
- Provides Quantitative Estimates for Events
- Evaluates the Value of Additional Information
- Mitigates Imprecision of Probability Phrases
- Ensures Mathematical Rules are Followed
- Forces the Consideration of Alternatives
- A Numerical Score / Range must be Defined

Subjective probability is a quantitative expression of an analyst's degree of belief in the truth of a statement relative to all other alternative possibilities.

PHASE2 © 2023 CHAPTER 10

Subjective Probability: Sample Table

Subjective Probability Table		
Term	Score	Range (Percentage)
Highly Probable	18-20	91 to 100
Probable	16-17	81 to 90
Highly Likely	11-15	71 to 80
Likely	8-10	61 to 70
Possible	5-7	41 to 60
Unlikely	4	31 to 40
Highly Unlikely	3	21 to 30
Improbable	2	11 to 20
Highly Improbable	1	1 to 10

Subjective probability does not determine a timeline for events; it attempts to allow analysts to predict an outcome by applying a percentage of probability.

PHASE2 © 2023 CHAPTER 10

Subjective Probability: Methodology (I)

APT33 (ELFIN TEAM) SUBJECTIVE PROBABILITY TABLE

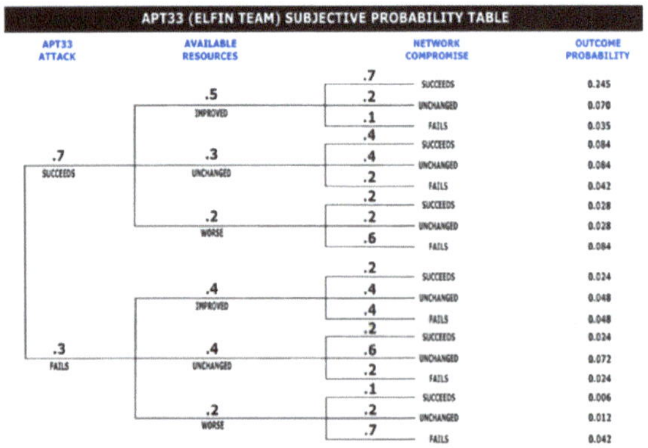

PHASE2 © 2023 CHAPTER 10

Subjective Probability: Methodology (II)

Threat Matrix: APT Comparative Analysis					
Risk Values 1-20	**Known TTPs** Documented	**APT 28** Fancy Bear	**APT 32** OceanLotus	**APT 33** Elfin Team	**APT 41** Wicked Panda
1-20	Targets Petrochemical and Energy Industries	0	0	20	1
1-20	Targets Regions of Organizational Operation	0	0	20	1
1-20	Capable of Executing Zero-Day Exploits	2	4	5	6
1-20	Uses Boot or Auto-Start Execution	15	15	10	10
1-20	Targets Non-Standard Ports and/or Protocols	15	15	10	1
1-20	Uses Encryption for C2 and Data Exfiltration	15	15	15	10
1-20	Targets Windows Servers and File Sharing Systems	15	15	15	10
1-20	Uses Spear Phishing to Compromise Accounts	20	20	20	7
Total Value 1-160		82	84	115	46

Small Number Documented	Moderate Number Documented	Significant Number Documented
Risk Values		
1-4 / **Low Probability**	5-10 / **Moderate Probability**	11-20 / **High Probability**

Chapter 10 Summary: Key Concepts

- Structured Analytic Techniques
- Sorting Methodology
- Matrices Methodology
- Event Mapping Methodology
- Event Tree Methodology
- Subjective Probability Methodology

Identifying and Exploiting Network Relationships

Relationship Modeling and Analysis (I)

- Requires Significant Time to Create
- The Model must be Constantly Maintained
- Continuous Maintenance Demands Effort
- - Referred to as "**Feeding the Beast**"
- Highly Effective in Analyzing Modern Issues
- The Graphical Displays are Powerful Tools
- - Persuading Customers to Accept Results

Relationships among entities – people, places, things, and events – are perhaps the most common and frequently used subject of intelligence modeling.

Relationship Modeling and Analysis (II)

- **Hierarchy Models** (Simple Tree Models)
- **Matrix Models** (Grids Reveal Relationships)
- **Link Models** (Complex Tree Structures)
- **Network Models** (Multiple Hierarchies)
- - *Social Network* (Human Relationship)
- - *Target Network* (Entity Relationships)

There are four levels of relationship models, each using increasing sophisticated analytic approaches: *hierarchy*, *matrix*, *link*, and *network* models.

The Netwar Model

- **Network** (aka: *System*)
- - Groups Forming a Unified Whole
- **Node**
- - Any Element of a System
- **Cell**
- - Formed Around a Specific Process
- **Link**
- - Physical or Functional Node Relationships

The *netwar model* of multidimensional conflict between opposing networks is increasing applicable to all intelligence and the network analysis tool of choice.

Link Models

- Display Relationships Among People
- Relationships Between People and Events
- Useful for Identifying Complex Relationships
- Minds cannot Assimilate all Collected Data
- **Goal**: Develop Precise and Valid Inferences
- - Used in Strategic and Operational Planning
- Software has Contributed Significantly
- - Display and Analyze Different Perspectives

Link modeling is routinely used in cybersecurity and law enforcement to identify Advanced Persistent Threats, Dark Web criminals, and terrorist organizations.

PHASE2 ADVANTAGE © 2023 CHAPTER 11

Network Models

- A Critical Tool for Intelligence Organizations
- Useful for Target Modeling (and Targeting)
- Planning Offensive Tactics and Strategies
- Relationships can be Positive or Negative
- Link Thickness used for Relationship Strength
- A Powerful Tool used in *Causal Modeling*
- Useful for the Creation of *Influence Nets*
- Color Coding can Indicate Types of Nodes

Network modeling and analysis has proven to be highly effective in intelligence organizations that focus primarily on both friendly and adversarial networks.

PHASE2 ADVANTAGE © 2023 CHAPTER 11

Network Model Types (I)

- Can Include Friendly or Allied Entities
- Can Include Neutral Entities to be Influenced
- - Convert to an Ally or Remain Neutral
- Can be Composed of Several Network Types
- - Threat, Social, Financial, or Organizational
- - Characterized by Nature of Relationships

There are numerous types of networks but for practical purposes, they are all *target networks*, even though the network is not necessarily hostile.

PHASE2 ADVANTAGE © 2023 CHAPTER 11

Network Model Types (II)

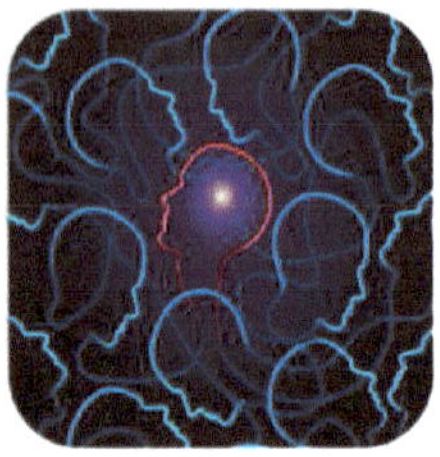

- **Functional Networks**
- - Formed for a Specific Purpose
- - Activities Based on Skills and Expertise
- **Family and Cultural Networks**
- - Culture, Language, Religion, or Ideology
- - Bonds Resulting from Geographic Ties
- **Virtual Networks**
- - Actors Seldom (or Never) Physically Meet

A target network can be any combination of types, and it is common for the network to include more than one form or type.

Manual Modeling

- **Understand the Environment**
- - Consider all 6 *PMESII* Variables
- - Consider all 6 *ASCOPE* Factors
- **Select or Create a Network Template**
- - Use Foundational Analytic Methods
- - Pattern, Link, or Social Network Analysis
- **Populate the Network Template**
- - Use the "*Snowballing*" Technique if Needed

When network modeling was first developed as an analytic methodology, the target models were created manually drawing circles and links on paper.

Computer-Assisted and Automated Modeling

- Attempt to Automate the Process Completely
- - Models are Created Directly from Raw Data
- Automated Network Models are Prevalent
- - Uses Data- and Text-Mining Techniques
- Analysts must Ensure to Evaluate Sources
- Analysts must Ensure to Validate Results
- Massive Datasets = Massive False Alarms

Although manual modeling is still used, commercially available network tools and software are available to help make the analyst's efforts more efficient.

Network Analysis

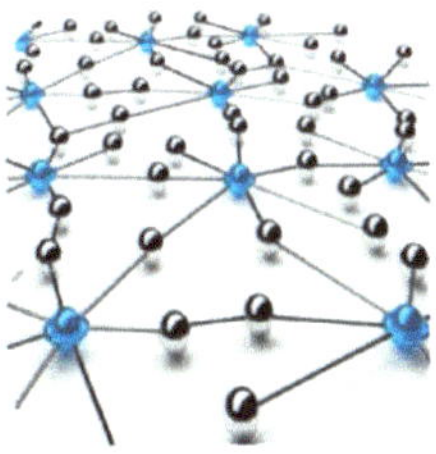

- **Network Pattern Analysis**
- - Identifies the *What, Where,* and *When*
- **Social Network Analysis**
- - Identifies the *Who*
- **Network Nodal Analysis**
- - Identifies the *How* and *Why*
- Identify the Function of Network Entities
- Identify the Importance of Network Entities

Network analysis involves addressing classic questions and placing the answers in a format the customer can understand and act upon (actionable intelligence).

Social Network Analysis

- Identifying and Assessing Relationships
- Links Show Relationships between Nodes
- Visual or Mathematical Nodal Relationships
- Used to Identify Key People in Networks
- Individuals are Referred to as "*Actors*"
- Two Fundamental Analytic Concepts are Used
- - **Centrality**
- - **Equivalence**

When intelligence analysts refer to network analysis, they are often referring to *social network analysis* (SNA) using platforms such as Twitter and Facebook.

Centrality

- Actors Occupy an Advantageous Position
- The Network Position Reveals Constraints
- Reveals the Ability to Constrain Others
- The Actor's Location must be Assessed
- Insight into Various Roles and Groupings
- - *Who are the Leaders and Connectors?*
- - *Where are Clusters and Who is in Them?*
- - *Who is in the Core of the Network?*

Centrality refers to the sources and distribution of power in a social structure, and the network perspective suggests power is derived from relationships.

Equivalence

- Useful for Strategic Targeting Missions
- Closely Related to the Concept of Centrality
- Nodes in the Network have Special Roles
- Network Nodes are Unique (no Equivalents)
- These Actors are Difficult to Replace
- Most Valuable Targets have **Centrality**
- Most Valuable Targets have no **Equivalents**

The disruptive effectiveness of removing one individual or a group of individuals from a network depends not only on centrality but on their uniqueness as well.

PHASE2 ADVANTAGE © 2023 CHAPTER 11

Target Value Analysis (TVA)

- **High-Value Targets (HVTs)**
- - Important to the Adversary's Leadership
- - Important for the Mission's Success
- **High-Payoff Targets (HPTs)**
- - A Critical High-Value Target
- - Loss will Degrade the Adversary's Success
- - The Loss will Support Friendly Success
- Both Targets have High Network Centrality

Target Value Analysis (TVA) is a method of identifying and prioritizing High-Value (HVTs) and High-Payoff Targets (HPTs) to determine which should be engaged.

PHASE2 ADVANTAGE © 2023 CHAPTER 11

Developing a High-Payoff Target List

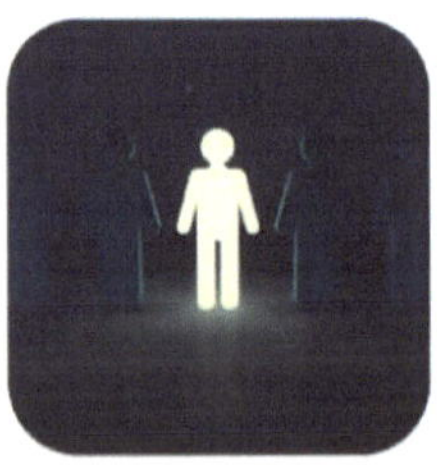

- **1**: Identify and Prioritize all HVTs
- **2**: Eliminate the Targets Beyond Reach
- **3**: Eliminate the Targets Beyond Capabilities
- **4**: Eliminate the Targets Beyond Timelines
- **5**: The Remaining HVTs Make the HPT List
- **6**: Coordinate IR Tasking for all HPTs
- **7**: Disseminate the List to Collection Teams

Engaging targets that the adversary can least afford to lose deprives them of their functionality and forces them to alter their established decision cycle.

PHASE2 ADVANTAGE © 2023 CHAPTER 11

Network Nodal Analysis

- May Involve Assessing Object Performance
- - Datacenters and/or Malware
- Cyber Threats can be Large and Complex
- Some Objects may be Critical Components
- Identify the most Critical Network Nodes
- - Requires Analysis of Individual Nodes
- - Understanding the Behavior of Nodes
- - Activities Taking Place within the Nodes

Nodes in a target network can include people, places, objects, and organizations, all of which can be treated as separate networks if required.

PHASE2 ADVANTAGE © 2023 CHAPTER 11

Threat Network Analysis (I)

- They are Opposed to Friendly Networks
- They may have State Backing
- They can Assume many Forms and Functions
- - **Insurgents**
- - **Violent Jihadists**
- - **International Criminal Organizations**

Threat networks have been defined as being "comprised of people, processes, and places – components that can be identified, targeted, and exploited."

PHASE2 ADVANTAGE © 2023 CHAPTER 11

Threat Network Analysis (II)

- Understand how the Network Operates
- Identify Actions to Exploit Vulnerabilities
- Some Threat Networks are Territorially Based
- Some Threat Networks are Transnational
- Have Access to *PMESII* Resources
- May Connect to Social Structures
- May Utilize Global Information Networks
- - Recruiting, Funding, and Information Ops

A premise of threat network modeling is that these networks have vulnerabilities that can be exploited once they are identified by intelligence analysts.

PHASE2 ADVANTAGE © 2023 CHAPTER 11

Threat Network Analysis (III)

- **C2 Structure** (Organizational)
- **Closeness** (Commonalities)
- **Expertise** (Knowledge, Skills, and Abilities)
- **Resources** (Financial and Public Support)
- **Adaptability** (Learn and Adjust Behaviors)
- **Sanctuary** (Safe Areas to Conduct Missions)

Analysts develop a detailed understanding of how threat networks function by identifying its elements, processes, and how the components interact.

Threat Network Analysis (IV)

- They may have Several Strengths
- - Ability to Adapt over Time
- - Can Blend into the Local Population
- - Can Quickly Replace Key Personnel Losses
- - Can Quickly Recruit New Members
- They are Typically Organized into Cells
- They may have Exploitable Weaknesses
- - Tend to Follow SOPs from Past Successes

Some threat networks must compete for resources, support, and territories, which requires communication that can be exposed and mapped by linking.

Target Network Analysis

- Entities in SNA are Typically People
- Target Network Analysis can be Anything
- Include Associated Locations and Activities
- **Goal**: Reveal Patterns and Future Targets
- Networks Typically Fall into Three Categories
- - **Threat Networks**
- - **Neutral Networks**
- - **Allied Networks**

Analysts usually apply an extension of social network analysis to all intelligence work, and SNA techniques work for almost all *target network analysis* missions.

Automating Network Analysis

- **Organization Risk Analyzer** (aka: *ORA*)
- - Can Group Network Nodes
- - Identify Patterns of Network Significance
- - Identify Key Players and Groups
- - Identify Potential Vulnerabilities
- - Model Spatial and Temporal Changes
- - Relies Heavily on Graphical Techniques
- - Can Store and Manipulate Massive Datasets

Target network analysis has become one of the principal tools for dealing with complex systems as a result of computer-based analytic methods and software.

PHASE2 ADVANTAGE © 2023 CHAPTER 11

Chapter 11 Summary: Key Concepts

- Relationship Modeling and Analysis
- Network Model Types
- Overview of Network Analysis
- Centrality and Equivalence
- Threat Network Analysis
- Target Network Analysis

PHASE2 ADVANTAGE © 2023 CHAPTER 11

Using Simulations for Intelligence Decisions

PHASE2 ADVANTAGE © 2023 CHAPTER 12

Overview of Simulation Modeling

- Impractical to Measure all Required Variables
- Used to Estimate Real-World Performance
- Estimates Made under Different Conditions
- Helps Customers Choose Among Alternatives
- - Determines Likely Outcomes of Actions
- Useful for Making Long-Range Forecasts
- - Exact Numerical Estimates not Needed

Simulation models are mathematical descriptions of the interrelationships believed to determine a system's behavior under different conditions.

Types of Simulations (I)

- **Deterministic**
- - A Spreadsheet with Rows and Cells
- - The Inputs are Fixed by Cell Numbers
- - Only One Solution Appears
- **Stochastic**
- - Uncertainty about Proper Inputs
- - Values are Assigned to Uncertain Inputs
- - Referred to as a *Monte Carlo* Simulation

The challenge of a ***stochastic simulation*** is to select the right uncertainties for the input variables which represent the correct probability distribution.

Types of Simulations (II)

- **Continuous Simulations**
- - The System Changes with Time
- **Discrete-Event Simulations**
- - System Changes States as Events Occur
- **Hybrid Simulations**
- A Combination of Continuous and Discrete

The simulation models typically used by intelligence analysts are classified as *continuous simulations*, *discrete-event simulations*, or a *hybrid* of the two.

Creating and Running a Simulation

- **1**: Formulate the Issue
- - Analytic Results: *Measures of Effectiveness*
- - Also Referred to as *"Figures of Merit"*
- **2**: Identify the Needed Input Information
- **3**: Select the Simulation Software
- **4**: Validate the Model
- - Proven and Validated Models are Preferred
- **5**: Run the Simulation and Interpret Results

Sophisticated simulation models can be complicated, and independent checks against real-world results are difficult to run on them.

Simulations Used in Intelligence Analysis

- **1**: Network Simulations
- **2**: Offensive Capability Simulations
- **3**: Weapons Systems Simulations
- **4**: Economic Simulations
- **5**: Geospatial Simulations
- **6**: Political and Social Simulations

Intelligence analysts use interactive simulation models extensively to assess the performance of adversarial threats, capabilities, and even financial economies.

Network Simulations

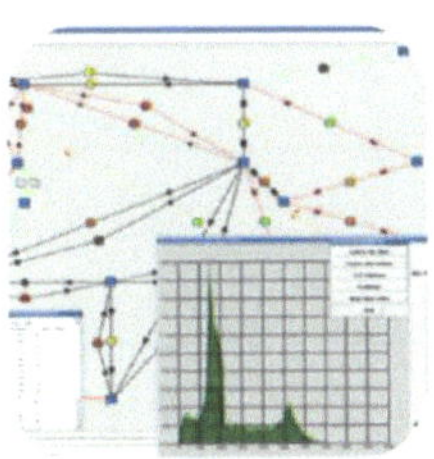

- Demonstrates Network Flows of Variables
- - **Money**
- - **Information**
- - **Technology**
- - **Disease**
- Assess Network Impact of Removing Nodes
- Assess most Important Nodes and Links
- Assess Estimated Network Recovery Times

Network simulations are a powerful tool used by intelligence analysts for modeling and analyzing threat, neutral, and allied networks.

Offensive Capability Simulations

- Used Primarily by Defense Organizations
- - Planning Systems Acquisition
- - Determining Force Mixes
- - Personnel Training (Technical and Combat)
- Incorporates Several Types of Models
- - Vulnerability and Cyber Effect Models
- - System Reliability Models
- - Force-on-Force Campaign Models

Also referred to as "war gaming," *offensive capability simulations* are closely related to offensive cybersecurity simulations and often incorporate them.

Weapons System Simulations

- Requires High-Speed Supercomputers
- Requires Complex Simulation Codes
- Many Simulation Models Already Exist
- Commercial Simulation Models are Available
- - Radars and Communication Systems
- - Satellite Orbits, Aircraft, and Shipping
- Details can Easily Overwhelm Analysts
- - Possible to Lose Sight of Main Objectives

Weapons systems simulations range in complexity from simulating performance of a single entity to that of elaborate interconnected systems.

Economic Simulations

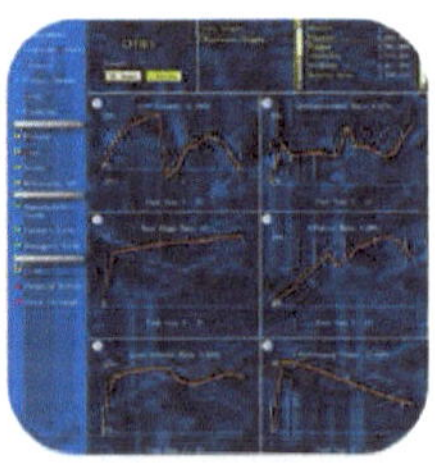

- A Quantitative Description of the System
- Incorporates Several Possible Hypotheses
- Describes how Economic Systems Function
- Used by Intelligence and Financial Agencies
- - Assess Trade and Balance of Payments
- - Assess Energy Production and Distribution
- Econometric Models Support Negotiations
- - Running Econometric Models is Preferred

Economic models (or *econometric models*) are used to simulate the performance of an entire economy or any segment of it such as competitive industries.

Geospatial Simulations

- Simulations can Take Several Forms
- Laser Radar and Synthetic Aperture Radar
- - Locations for Pandemic Treatment Centers
- - Terrain Mapping and Flood Models
- Need Geographic Coordinates for Nodes
- Combine Network and Geospatial Simulations
- - Used for Predicting Population Movements

One of the most powerful applications of *geospatial simulations* comes from combining them with models such as *Operational Requirements Analysis* (ORA).

PHASE2 ADVANTAGE © 2023 CHAPTER 12

Political and Social Simulations

- Simulations Receive Data from many Sources
- - Historical Patters
- - Web Activity
- - Social Media
- **1**: Warn of Possible **Disasters**
- **2**: Warn of Possible **Disease Activity**
- **3**: Warn of Possible **Trends in Crime**
- **4**: Warn of Possible **Geopolitical Events**

Political and social simulations have been developed to analyze the interactions of political parties, diplomatic initiatives, and clashes across cultural fault lines.

PHASE2 ADVANTAGE © 2023 CHAPTER 12

Decision Modeling and Simulation

- **1**: Rational Models
- **2**: Administrative Models
- **3**: Cultural Models
- **4**: Emotional Models
- **5**: Collective Decision-Making Models
- **6**: Game Theory

Decision modeling and simulation analyzes the personality, problem-solving styles, values, goals, behaviors, and environments of individual leaders.

PHASE2 ADVANTAGE © 2023 CHAPTER 12

Rational Models

- The Choices Involved are Identified
- Values are Assigned for Possible Outcomes
- - Referred to as a *"Cost-Benefit Analysis"*
- The Probability of Outcomes is Expressed
- The Rational Model is Based on Two Concepts
- **1**: *Optimal Choice*
- **2**: *Expected Utility Theory* (EUT)
- - Start with Rational Decisions and EUT

Rational decision-making requires the systematic evaluation of costs or benefits accruing to courses of action that may be taken by an individual.

PHASE2 © 2023 CHAPTER 12

Administrative Model

- Effort to Find Optimum Decision is not Made
- Complexity of the Problem Dissuades Effort
- A Tendency to Select Several Outcomes
- - Possible Outcomes must be "Good Enough"
- Discards the Three Assumptions of *EUT*
- **1**: People have Incomplete Information
- **2**: People have Time Pressures
- **3**: People Beset with Conflicting Preferences

The tendency of leaders to adopt suboptimal choices leads to a variant of the rational decision-making simulation referred to as the *administrative model*.

PHASE2 © 2023 CHAPTER 12

Cultural Models

- Must Understand Elements of a Culture
- - Training Youth for Adult Roles
- - How it Defines Importance in Life
- Culture Defines Ethical Norms of Collective
- It Dictates Values and Constrains Decisions
- - A Constraining Social / Environmental Force
- **Risk**: The "Irrational Behavior" Trap
- **Risk**: Imposing "Mirror Imaging" in Analysis

Behavior cannot be predicted with any degree of analytical confidence without the analyst putting it in the decision maker's social and cultural context.

PHASE2 © 2023 CHAPTER 12

Emotional Models

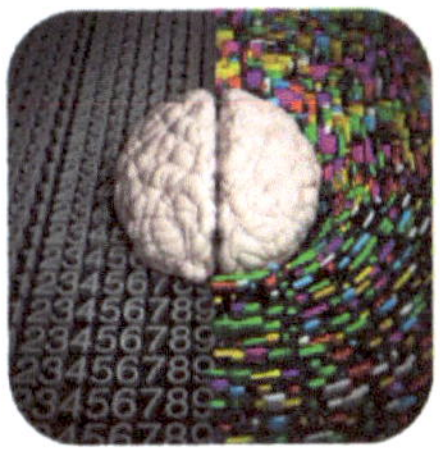

- Exciting and Challenging Things Attract Us
- Pride and Revenge are also Motivators
- Emotional Aspects must be Addressed
- Methods Exist for Personality Profiling
- - Idiosyncrasies and Predisposition Prevail
- - Calculated Assessments are Secondary
- **Leader**: College Linebacker vs. Chess Club
- Profile Models can Assess Likely Decisions

An important aspect of the decision process to consider when analyzing behavior is the emotional one, a dominant component in many influential leaders.

PHASE2 ADVANTAGE © 2023 CHAPTER 12

Collective Decision-Making Models

- Easier to Predict Groups than Individuals
- Primary Decision-Makers must be Identified
- The Model has Three Dominant Features
- **1**: Power is Decentralized
- **2**: Multiple Competing Goals Exist
- **3**: Decisions Result from Bargaining
- A Complex Process of Conflict Resolution
- - Decisions are Products of Compromises

Decisions are often made by a group and such cases require that intelligence analysts employ a *collective decision-making model* for their predictions.

PHASE2 ADVANTAGE © 2023 CHAPTER 12

Game Theory

- The Game Assumes Two or More Players
- Each Determines how the Other will Respond
- Each Player Determines Their Own Response
- - Must Estimate Moves of other Players
- - How Initial Decision Result in Game's End
- Player Analyzes all Available Information
- Players Identify an Initial Preferred Decision
- - Understand the Decision Process of Others

Game theory analyzes the decision process of two or more parties in conflict or cooperation and is valuable in simulation for its use of mathematical equations.

PHASE2 ADVANTAGE © 2023 CHAPTER 12

Chapter 12 Summary: Key Concepts

- Overview of Simulation Modeling
- Types of Simulations
- Creating and Running a Simulation
- Simulations Used in Intelligence Analysis
- Decision Modeling and Simulation
- Game Theory and Decision Response

PHASE 2 ADVANTAGE © 2023 CHAPTER 12

Making Decisions Based on Potential Outcomes

PHASE 2 ADVANTAGE © 2023 CHAPTER 13

The Purpose of Outcome Scenarios (I)

- Helps all Involved to Anticipate the Future
- A Better Response to Subsequent Events
- Helps Communicate Non-Existent Situations
- Conveys the Results of Decisions to Action
- - Difficult to Quantify Complex Situations
- Strategic, Operational, and Tactical Uses
- - Presents a Set of Alternative Scenarios
- - Reflects a Range of Plausible Outcomes

The purpose of scenarios is to highlight significant forces that may change the future, making the forces visible and easier for an analyst to recognize.

PHASE 2 ADVANTAGE © 2023 CHAPTER 13

The Purpose of Outcome Scenarios (II)

- Predicting the Future in Detail is not Possible
- Details are Controlled by Divergent Factors
- Trends are Controlled by Convergent Factors
- - Allows for "Most Likely" Outcome Scenarios
- - Indicators Reveal "Most Likely" Outcomes
- Track the Indicators for Specific Scenarios

In an alternative future depicted by a scenario, a customer should be able to make decisions and develop strategies using outcomes developed by the analyst.

The Role of Outcome Scenarios

- Convey Possible Courses of Events
- - Some Events may be Difficult to Grasp
- - Far Removed from Anything Envisioned
- Anticipatory Analysis Results in Scenarios
- - Describes the Future State of the Target
- Scenarios are Typically Presented as a Story
- Used for Planning and Decision Making
- Usually Presented with Augmented Reality

Outcome scenarios present several scenarios for an uncertain future about what potential events may lie ahead, each one modeling a plausible outcome.

Types of Scenarios

- **1**: *Demonstration Scenarios*
- **2**: *Driving-Force Scenarios*
- **3**: *System-Change Scenarios*
- **4**: *Slice-of-Time Scenarios*
- - A "Snapshot" of the Final Future Scene

Analysts use four basic types of outcome scenarios: *demonstration scenarios*, *driving-force scenarios*, *system-change scenarios*, and *slice-of-time scenarios*.

Demonstration Scenarios

- A High-Impact / Low-Probability Product
- Identifies Triggering Events along the Path
- Represent Decision or Occurrence Points
- - These Points may Shape the Outcomes
- Identify Later Occurrences of Events
- - May Indicate Development of a Scenario
- Focus Attention on the Triggering Events
- - May Change the High-Impact Outcome

When analysts create a *demonstration scenario*, they identify the high-impact outcomes and then the plausible paths that may lead to that result.

Driving-Force Scenarios

- Create Multiple Driving-Force Scenarios
- Identify the Key Uncertain Factors and Forces
- - Key Forces will Shape Potential Outcomes
- Identify Alternative Change Levels for Each
- Develop a Matrix or Tree to Visualize Results
- - Shows Outcomes for Assumed Changes
- Describes the Effects of Changing Forces
- - Alternative Force Levels Remain Constant

A *driving-force scenario* is an estimative approach for projection in which major forces acting on a target are examined to determine how they are changing.

System-Change Scenarios

- Considers all Significant PMESII Variables
- Accounts for Changes in Existing Forces
- Accounts for Introduction of New Forces
- Scenario Relies on *Cross-Impact Analysis*
- - Identifies Interactions Among Events
- - Interactions Develop Alternative Outcomes
- An Especially Difficult Scenario to Develop
- - Examination of Interrelationships of Forces

A *system-change scenario* is a variant of the driving force scenario and accounts for the fact that alternative force levels remain constant in the estimative model.

Slice-of-Time Scenarios

- Resembles the Demonstration Scenario
- - An Endpoint Outcome is Described
- - No Pathway to the Endpoint is Provided
- Not Typically Useful in Actionable Intelligence
- - No Focus on Forces Leading to Outcome
- - Provides Minimal Indicators to Monitor
- Results are Often Seen in Policy Papers
- - Papers Define a Specific Environment

A *slice-of-time scenario* starts at a time in the future in which a synthesis of explicit or assumed forces has already created a specific outcome.

Scenario Perspectives

- The Scenario Usually Begins in the Present
- The Scenario Unfolds to a Future Time
- Extrapolations are **Momentum Scenarios**
- - They are Dominated by Inertia
- - No Opposing Forces Arise to Change Trend
- - A Poor Choice for Long-Term Assessments
- Customers Prefer **Normative Scenarios**
- **Goal**: Develop a Course of Action

Scenarios are models, and like the models utilized by intelligence analysts which were discussed in previous domains, they can be either *descriptive* or *normative*.

PHASE2 ADVANTAGE © 2023 CHAPTER 13

How to Construct Scenarios

- **1**: Define the Issue and Target Framework
- **2**: Identify and Rank the Driving Forces
- **3**: Select the Scenario Logic
- **4**: Flesh Out the Scenario
- **5**: Draw Out the Implications
- **6**: Identify and Monitor Indicators

Scenario planning is a standardized and systematic methodology that derives from well-known modeling, simulation, and gaming theories used in the past.

PHASE2 ADVANTAGE © 2023 CHAPTER 13

1: Define the Issue and Target Framework

- Ask Relevant Preliminary Questions
- Define the Issue or Problem to be Solved
- Conduct a Key Assumptions Check
- Select the Relevant Issue Categories
- Decompose the Issue and Sub-Issues
- Select the Target Model Type to be Used
- Create the Target Framework for the Issue
- Develop Alternative Target Frameworks

For most intelligence issues, there are typically several customers interested in receiving intelligence who have different interests or questions to be answered.

PHASE2 ADVANTAGE © 2023 CHAPTER 13

2: Identify and Rank the Driving Forces

- Isolate the Known Driving Forces
- Identify Forces Critical to the Outcome
- - Some Driving Forces Affect Everyone
- Identify any Present Asymmetric Forces
- Rank the Driving Forces by **Importance**
- Rank the Driving Forces by **Uncertainty**

Because scenarios are a way of understanding the dynamics driving the future, analysts attempt to identify the primary driving forces shaping the present.

PHASE2 ADVANTAGE © 2023 CHAPTER 13

3: Select the Scenario Logic

- This is Probably the most Important Step
- Derived from Fundamental Differences
- Work with Issues, Targets, and Forces
- Reshape and Reframe the Components
- Draw Out Their Less Obvious Elements
- Review Components for Emerging Patterns
- A Few Forces will Differentiate the Outcome
- - They are Referred to as **Scenario Drivers**

Analysts want to finish with a few scenarios – typically four or fewer – that provide their customer with the basis for making the best decision possible.

PHASE2 ADVANTAGE © 2023 CHAPTER 13

4: Flesh Out the Scenario

- Expanding the Detail of Scenario Elements
- Three Techniques are Commonly Used
- - **Case-Based Models**
- - **Contextual Mapping**
- - **Cross-Impact Analysis**

Analysts now need to return to step two of this process, "*Identify and Rank the Driving Forces*," and use those elements to further flesh out the scenario.

Case-Based Models

- Using Old Experiences to Solve New Issues
- - Analysts Consider a Previous Situation
- - The Situation is Similar to the Current One
- - Use the Old Situation to Solve the New One
- Old Solutions Adapted to Meet New Demands
- Reasoning from Precedent used to Interpret
- Reusing Knowledge (Duplicating the Model)

Case-based models are the foundation for a type of analysis called case-based reasoning which uses reasoning by analogy or reasoning by history.

Contextual Mapping

- Useful for Forcing a Fresh Perspective
- Useful for Predicting Actual Developments
- Requires Experienced Analytical Experts
- - Familiar with Method and Topic of Inquiry
- Primarily Used in Technological Forecasting
- Future Developments = External Forces
- Used as a Graphical Display with Timelines
- Interconnecting Paths Lead to Developments

The technique of *contextual mapping* is used to identify plausible sequences of development in a field and relate them to developments in a different field.

Cross-Impact Analysis

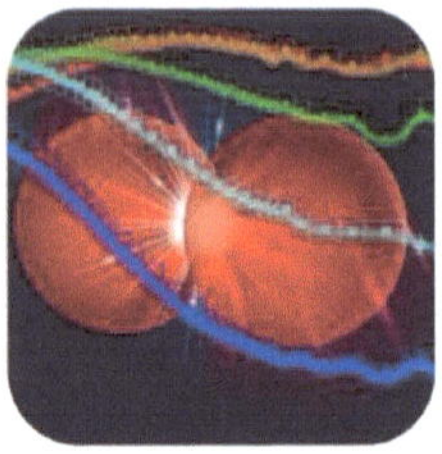

- Used to Support System-Change Scenarios
- Shows Interactions Among Developments
- This Type of Analysis is Used Extensively
- - Model the Interaction of Future Events
- - Model the Interaction of Future Trends

Cross-impact analysis **specifies how one event will influence the timing, model of impact, and likelihood of another event in a different but associated field.**

PHASE2 ADVANTAGE © 2023 CHAPTER 13

5: Draw Out the Implications

- Tactical, Operational, and Strategic Impacts
- Consider Laws of Unintended Consequence
- Everyone must Consider Possible Outcomes
- - Primary Customers of Intelligence
- - Policy Executives and Decision-Makers
- Options must be Developed Accordingly

Having built out a set of potential future scenarios and outcomes, the analyst must examine each of them carefully in the context of the original question.

PHASE2 ADVANTAGE © 2023 CHAPTER 13

6: Identify and Monitor Indicators

- Look for the Leading Indicators
- Identify the Scenario Taking Place
- - Can be a Combination of Scenarios
- - Can Indicate Scenarios **not** Taking Place
- Monitoring may Involve Watching Trends
- Select Options to Focus Customer Decisions
- Identify Outcomes Requiring Preparation
- Aka: "**Indicators or Signposts of Change**"

The job of the intelligence analyst (and customer) becomes one of *identifying and monitoring indicators* once the scenarios have been completed.

PHASE2 ADVANTAGE © 2023 CHAPTER 13

Devil's Advocates

- The Role Seeks to Directly Engage Others
- - An Argumentative Discussion Process
- Attempts to Test the **Validity** of the Product
- Attempts to Test the **Quality** of the Product
- - Identify Weaknesses in its *Logic*
- - Identify Weaknesses in its *Assumptions*
- - Identify Weaknesses in its *Conclusions*
- Involves Pre-Event Preparation and Effort

On an intelligence team, a *devil's advocate* is someone who takes a position he or she does not necessarily agree with to provoke a debate.

PHASE2 ADVANTAGE © 2023 CHAPTER 13

Overview of Red Team Analysis (I)

- Used to Criticize Intelligence Assessments
- Employs Knowledge of Adversary's Culture
- Employs Knowledge of Adversary's Practices
- - Requires SME Knowledge of the Adversary
- Red Team Analysis is **not** Role-Playing
- - Not Submerged in the Adversarial Role
- Can Help Detect Errors in Measurements
- Provides a Reduction of Cognitive Biases

Red Teams are a form of challenge analysis used by intelligence teams that adopt the mindset and mentality of the adversary to become the "enemy."

PHASE2 ADVANTAGE © 2023 CHAPTER 13

Overview of Red Team Analysis (II)

- Address Mindsets of how Adversaries Think
- Address Beliefs of what Adversary's Needs
- Address Means to Achieve Their Objectives
- Used at the End of the Assessment Process
- Used in a Diverse Set of Scenario Contexts
- Users must Believe in Their Effectiveness
- - Teams must be Willing to Accept Findings
- Leadership must Support Challenge Analysis

It has been historically difficult for analysts to accept the Red Team findings of challenge analysis and for leaders to afford teams the time to fix problems.

PHASE2 ADVANTAGE © 2023 CHAPTER 13

Red Teams: Understanding Adversaries

- **Different Beliefs**
- - A Radically Different Model of the World
- **Different Objectives**
- - Instability, Attention, Ideological Expansion
- **Different Capabilities**
- - Less Concern about Internal Opposition
- **Different Moral Constraints**
- - Objectives vs. Moral Costs / Human Rights

Red teaming is often used to analyze an adversary's decision-making, assessing likely strategies and tactics, and predicting the outcomes of thought processes.

Chapter 13 Summary: Key Concepts

- The Purpose of Outcome Scenarios
- The Role of Outcome Scenarios
- Types of Scenarios
- Scenario Perspectives
- How to Construct Scenarios
- Using Red Team Analysis

Considerations for Creating an Intelligence Team

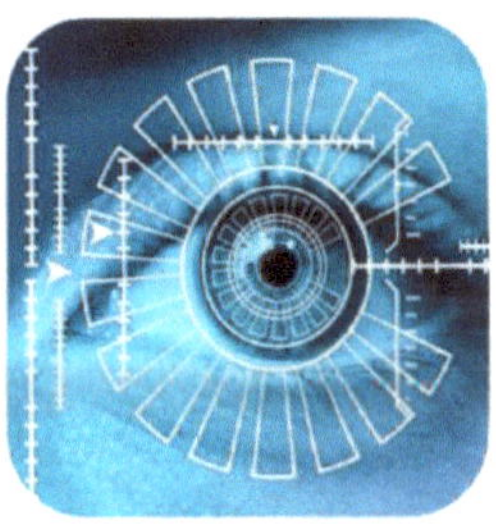

The Threat Intelligence Analysis Team

- Provides Customers a Decision Advantage
- Provides Customers a Competitive Advantage
- Influences the Adversary's Decision Cycle
- Provides Understanding of Complex Issues
- Contributes to Overall Operational Efficiency
- Contributes to an Enhanced Security Posture
- Supports the Organization's Strategic Goals

Many organizations attempt to create a threat intelligence analysis team without a comprehensive plan or an understanding of the needs driving the initiative.

Approaches to Building an Intelligence Team

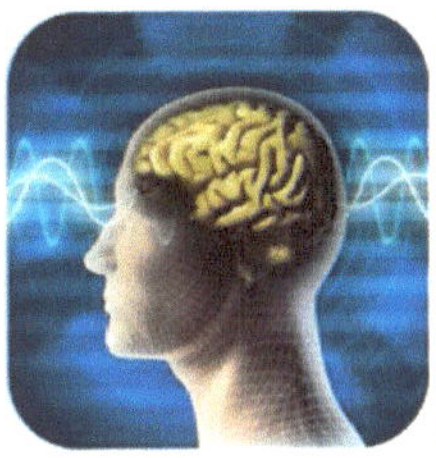

- Involves Cognitive and Social Processes
- Analysis is a **Cognitive Activity**
- - A System of Heuristics and Conclusions
- Analysis is a **Social Activity**
- - Analysts Draw on the Expertise of Teams
- "Lone Wolf" Analysts are the Exception

Although analysts draw on technological aids and inputs, it is ultimately the human brain that organizes and interprets data to generate assessments.

The Cognitive Perspective

- Puts Individual Analyst at the Center Stage
- Organizes and Encourages Individual Work
- Supports Excellent Individual Performance
- Tends to Select Highly Talented Analysts
- Provides Analysts with Advanced Training
- Provides Sophisticated Technical Support
- Individual Analysts are Held Accountable
- Aka: **Co-Acting Groups**

The different perspectives on analytic work (cognitive and social) have several significant implications for the design and leadership of intelligence teams.

The Social Perspective

- Members are Collectively Responsible
- Projects Tend to be Larger in Size
- Projects Tend to be Potentially Significant
- Team Members have Specific Expertise
- Specialized Team Roles Evolve over Time
- The Entire Team Produces the Product
- The Entire Team is Held Accountable
- Aka: **Interdependent Work Teams**

The *social perspective* focuses on the importance of collegial interactions in competently assessing data and managing relationships with all those involved.

 CHAPTER 14

Co-Acting Groups vs. Work Teams

- *Is the group a **real** team?*
- *Does it have a compelling **purpose**?*
- *Does its **structure** facilitate teamwork?*
- *Is its organizational **context** supportive?*
- *Does it have access to **competent** coaching?*

Research and experience has identified five conditions which may increase the chances (but do not guarantee) that a Work Team will perform effectively.

 CHAPTER 14

8 Shared Factors for High Performance Teams

- **1**: A Clear and Compelling Direction
- **2**: The Right Size to Perform the Task
- **3**: Meaningful Work Assignments
- **4**: Regularly Receive Trustworthy Feedback
- **5**: Clear Norms about Acceptable Behaviors
- **6**: A Supportive Organizational Context
- **7**: Receive Recognition for Performance
- **8**: Get Coaching, Teaching, and Experience

Whether co-acting or interdependent, the best analytic teams are stable and bounded, with members interdependent for a particular shared purpose.

 CHAPTER 14

Team Manager vs. Peer Coaching

- Managers are Responsible for the Product
- - Attentions are Focused on Outputs
- Focus on Running External Interference
- - Dealing with Resources and Roadblocks
- Focus on Coaching Individual Members
- - Personal and Performance Issues
- Focus on Coaching the Team as a Team
- - Peer Coaching Remains a Prominent Force

Members of teams that are well designed experience stronger impulses to teach and assist one another than do members of co-acting groups.

PHASE2 ADVANTAGE CHAPTER 14

The Quality of Coaching in Teams

- Coaching Requires a Level of Skill
- - Not all Managers are Suitable Coaches
- Bad Coaching is Worse than no Coaching
- Insufficient to Ask for Additional Coaching
- Training in Coaching Skills is Required
- - Ensure Provided Coaching is Well-Timed
- - Ensure Coaching is Competently Delivered
- Ineffectual if the Team is Poorly Composed

Spending time on coaching intelligence teams does not guarantee improvement in the individual analyst's or the team's overall performance and behavior.

PHASE2 ADVANTAGE CHAPTER 14

Options for Intelligence Team Structures

- **Individual-Centric Teams**
- - Surgical Teams
- - Co-Acting Groups
- **Interdependent Work Team**
- - Face-to-Face
- - Virtual and Distributed
- **Dynamic Teams**
- - Sand Dune Teams

Experienced managers consider all options prior to settling on the one that is the most appropriate and feasible for assignment and the organizational context.

PHASE2 ADVANTAGE CHAPTER 14

Key Roles and Responsibilities

- The **4** Primary Roles in Intelligence Teams:
- **1**: Intelligence Team Manager
- **2**: Intelligence Collection Manager
- **3**: Fusion Intelligence Analyst
- **4**: All-Source Intelligence Analyst

As more organizations pursue intelligence as a critical security function, many are choosing to create dedicated teams with multidisciplinary capabilities.

Intelligence Team Manager Responsibilities

- Develop and Prepare Intelligence Strategies
- Validate the Quality of Intelligence Products
- Ensures Seamless Integration of Disciplines
- Develop, Prioritize, and Assign IR Taskings
- Oversee Reporting and Compliance Needs
- Mentor, Train, and Develop Team Members
- Manage Timely and Effective Dissemination

The *team manager* accepts responsibility for all aspects of the intelligence effort and interacts with both internal and external personnel and organizations.

Intelligence Team Manager Capabilities

- **7** Key Qualities of Effective Team Managers
- - Can Push Difficult Actions to Completion
- - Can Effectively Resolve Disagreements
- - Understands Intelligence and Technology
- - Has the Ability to Focus on Execution
- - Understands Nuances of Internal Politics
- - Has a Proven Record of Building Support
- - Has Skills to Communicate in all Disciplines

Team managers are required to develop innovative approaches to resolve difficult issues that can significantly impact all levels of the intelligence effort.

Intelligence Collection Manager Responsibilities

- Analyze Information Collection Requirements
- Evaluate Available Assets and Capabilities
- Recommend Taskings for Collection Assets
- Submit Requests for Collection Support
- Manage Timely Collection and Reporting
- Ensure Information is Relevant and Validated
- Assess Effectiveness of the Collection Efforts
- Coordinate Dissemination to Analytical Teams

Information collection is an activity that synchronizes the planning and employment of assets in direct support of current and future operations.

Intelligence Collection Manager Capabilities

- Ability to Prioritize and Focus on Execution
- Communicate with Multidisciplinary Teams
- Exceptionally Attention and Detail Oriented
- Understand Inputs from Multiple Sources
- Effectively Summarize Complex Issues
- Possess Highly Developed Reasoning Skills
- Excellent Written and Oral Presentation Skills

Conducting successful information collection requires thorough and creative planning, aggressive execution, and adjustments based on the situation.

Fusion Intelligence Analyst Responsibilities

- Determine the Utility of Collected Data
- Identify and Resolve Intelligence Gaps
- Communicate Collector and Analyst Needs
- Refine Individual Collection Processes
- Present Final Intelligence to Customers
- Write *Intelligence Information Reports* (IIRs)
- Educate Teams in Doctrinal Methodologies

The *fusion analyst* synthesizes data collected from intelligence disciplines, validates data for accuracy, and converts it into a finished intelligence product.

All-Source Intelligence Analyst Responsibilities

- Clearly Articulate Intelligence Requirements
- Formulate Research Questions and Variables
- Conduct In-Depth Research and Analysis
- Identify Threat Decision-Making Processes
- Identify Intelligence Gaps and Shortfalls
- Accurately Source all Intelligence Data
- Confidently Convey Intelligence Concepts
- Maintain Well-Organized Record Systems

All-source analysts analyze data from multiple sources, responds to information requests, and submits intelligence requirements in support of operations.

Intelligence Analyst General Characteristics

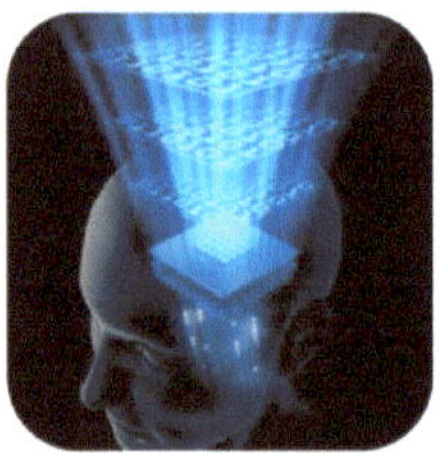

- Well-Disciplined and Highly Organized
- Can Communicate with Clarity and Credibility
- Identify Relevant Patterns in Chaotic Noise
- Utilize Patient and Methodical Approaches
- Can Holistically Visualize at Strategic Levels
- Can be Flexible and Adapt to Change
- Willing to Learn from Previous Mistakes

Intelligence analysts can look at dissimilar objects, thoughts, or actions and see in their minds eye patterns that untrained people would otherwise overlook.

Chapter 14 Summary: Key Concepts

- The Threat Intelligence Analysis Team
- Approaches to Building an Intelligence Team
- Cognitive and Social Perspectives
- Co-Acting Groups vs. Work Teams
- Options for Intelligence Team Structures
- Key Roles and Responsibilities

CHAPTERS 01 – 14

Course Appendix

Matrices and Charts

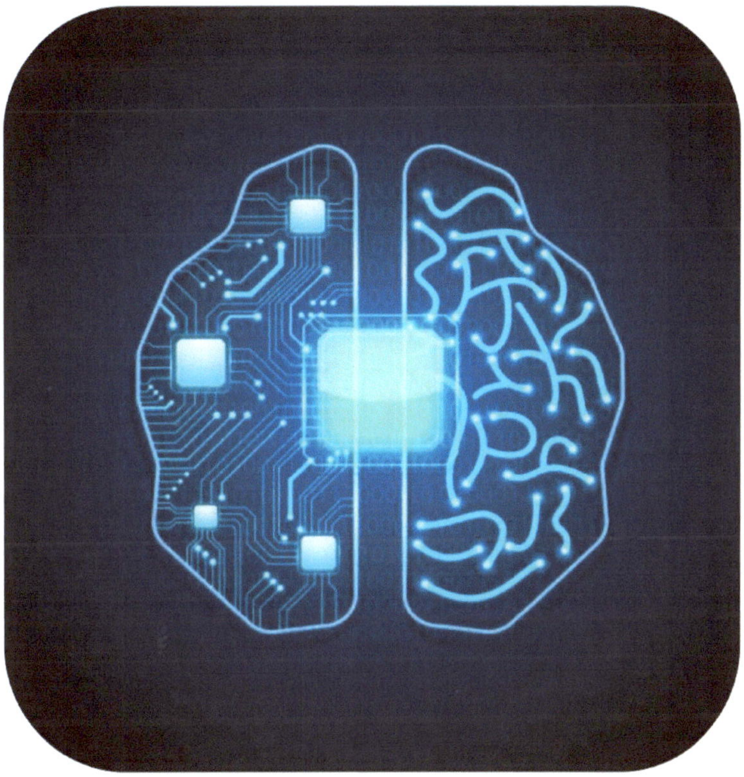

NOTICE TO STUDENTS:

This section of the course workbook contains all the matrices and charts presented in this course. You are encouraged to use these enlarged and detailed graphics taken from the lecture slides as memory aids during the certification exam.

Image 1.1: Final Analysis: Intelligence Product

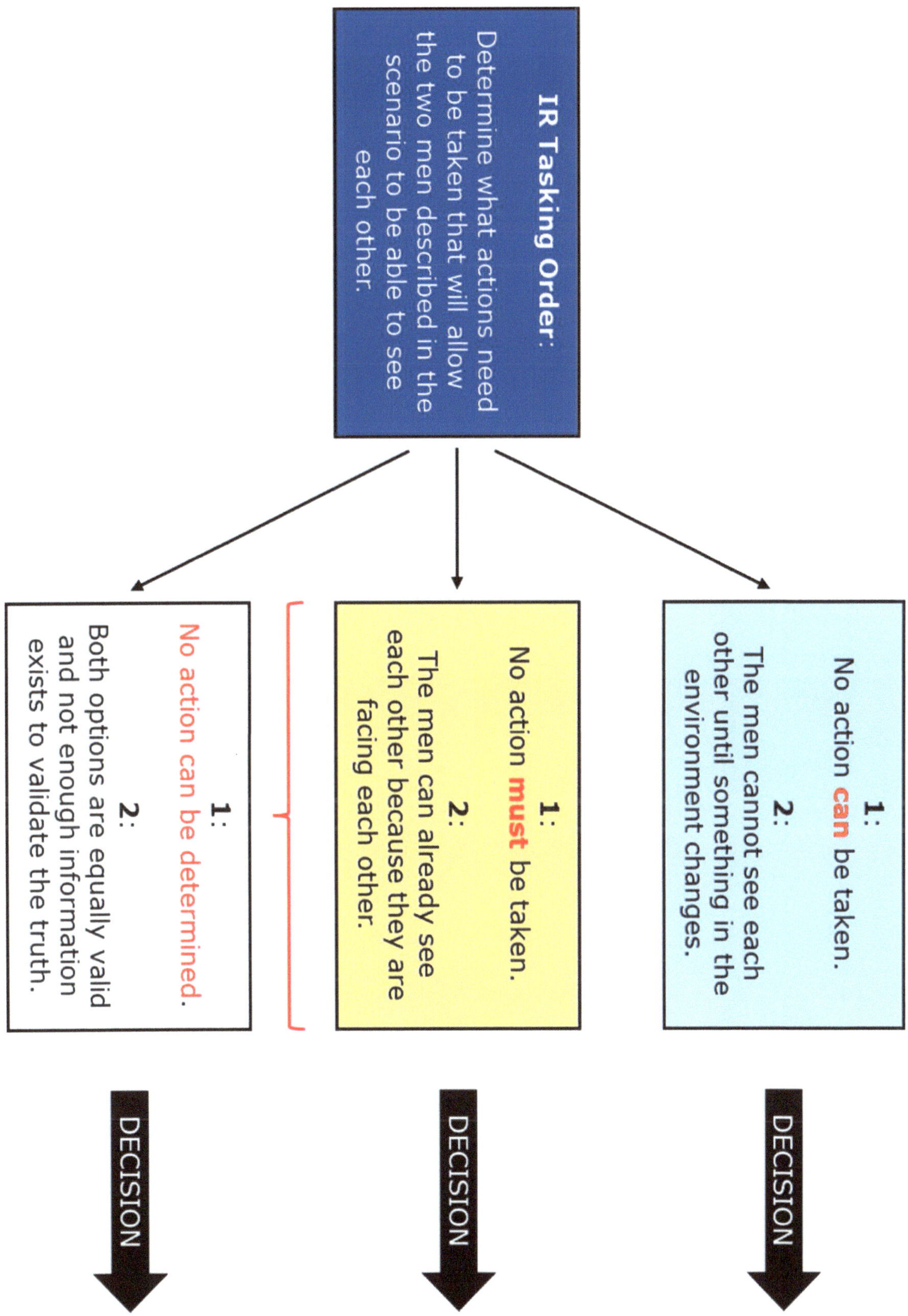

Image 2.1: Final Analysis: A Risky Jump

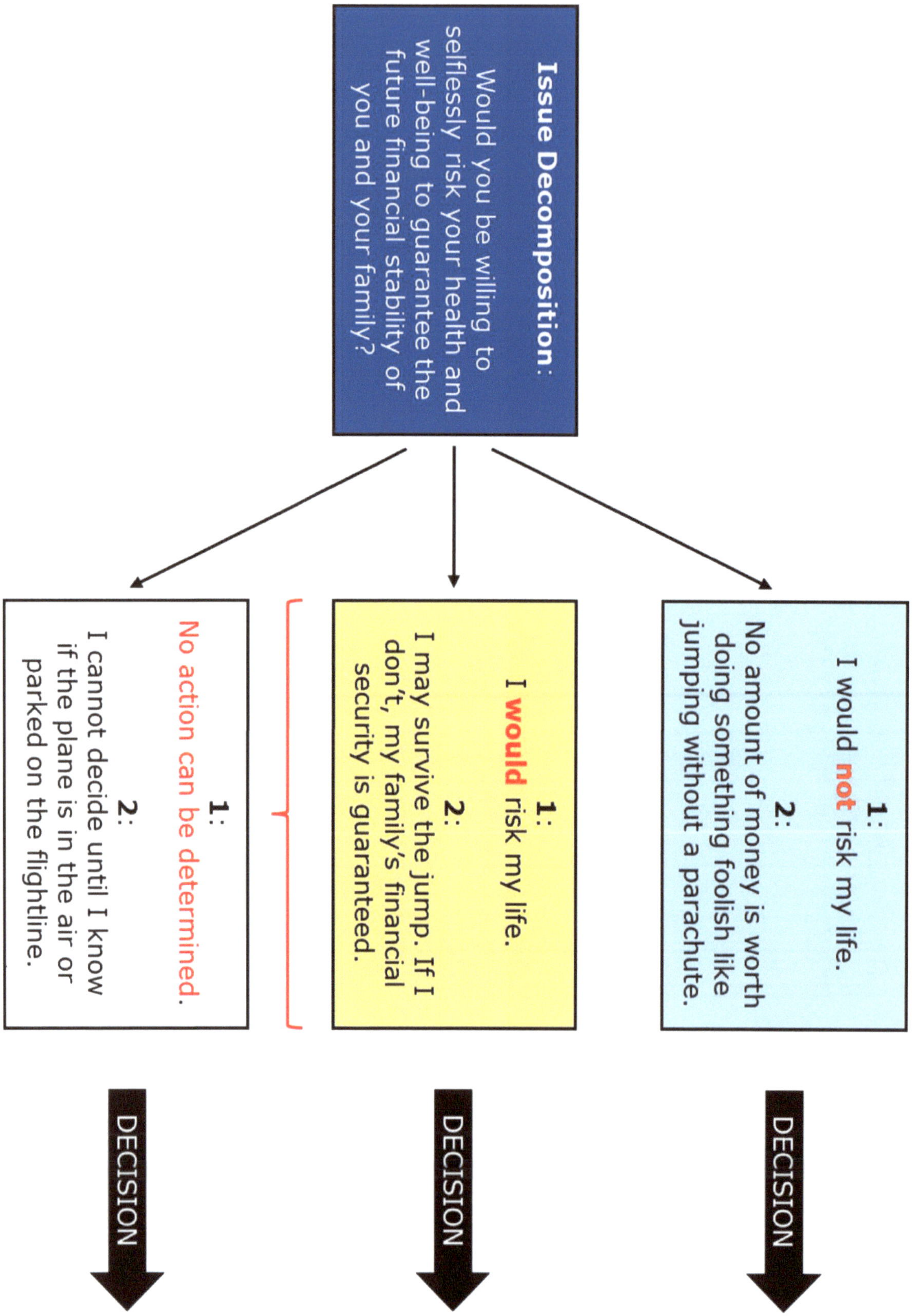

Image 4.1: Indicators of Falsified Information (I)

EMPLOYEE "A"

Employer Information

Employer:	Geek Tech Computer Repair
Address:	438 Mooreland Avenue NE
City, State, Zip Code:	Marietta, Georgia 30062
Telephone:	(770) 794-9390

Applicant Information

First Name:	Nicholas
Last Name:	Ferron
Address:	829 Jefferson Street NW
City, State, Zip Code:	Atlanta, Georgia 30308
Telephone:	(404) 506-1923
Date of Birth:	11/15/1977
Social (SSN):	157-43-1560
Driver License:	(GA) 047829025

EMPLOYEE "B"

Employer Information

Employer:	Savannah Bytes Programming
Address:	124 Stevens Court SW
City, State, Zip Code:	Savannah, Georgia 30124
Telephone:	(912) 487-5050

Applicant Information

First Name:	Michael
Last Name:	Guifanno
Address:	1 Riverside Drive, Apt. 24a
City, State, Zip Code:	Rincon, Georgia 31247
Telephone:	(912) 493-3446
Date of Birth:	01/24/1978
Social (SSN):	250-12-4693
Driver License:	(GA) 030612477

Image 4.2: Indicators of Falsified Information (II)

EMPLOYEE "**A**"

Employer Information

Employer:	Geek Tech Computer Repair
Address:	438 Mooreland Avenue NE
City, State, Zip Code:	Marietta, Georgia 30062
Telephone:	(770) 794-9390

Applicant Information

First Name:	Nicholas
Last Name:	Ferron
Address:	829 Jefferson Street NW
City, State, Zip Code:	Atlanta, Georgia 30308
Telephone:	(404) 506-1923
Date of Birth:	11/15/1977
Social (SSN):	157-43-1560
Driver License:	(GA) 047829025

EMPLOYEE "**B**"

Employer Information

Employer:	Savannah Bytes Programming
Address:	**124** Stevens Court SW
City, State, Zip Code:	Savannah, Georgia 30**124**
Telephone:	(9**12**) **4**87-5050

Applicant Information

First Name:	Michael
Last Name:	Guifanno
Address:	**1** Riverside Drive, Apt. **24**a
City, State, Zip Code:	Rincon, Georgia 3**124**7
Telephone:	(9**12**) **4**93-3446
Date of Birth:	0**1**/**24**/1978
Social (SSN):	250-**12**-**4**693
Driver License:	(GA) 0306**124**77

Image 5.1: Final Analysis: The Generals Disagree

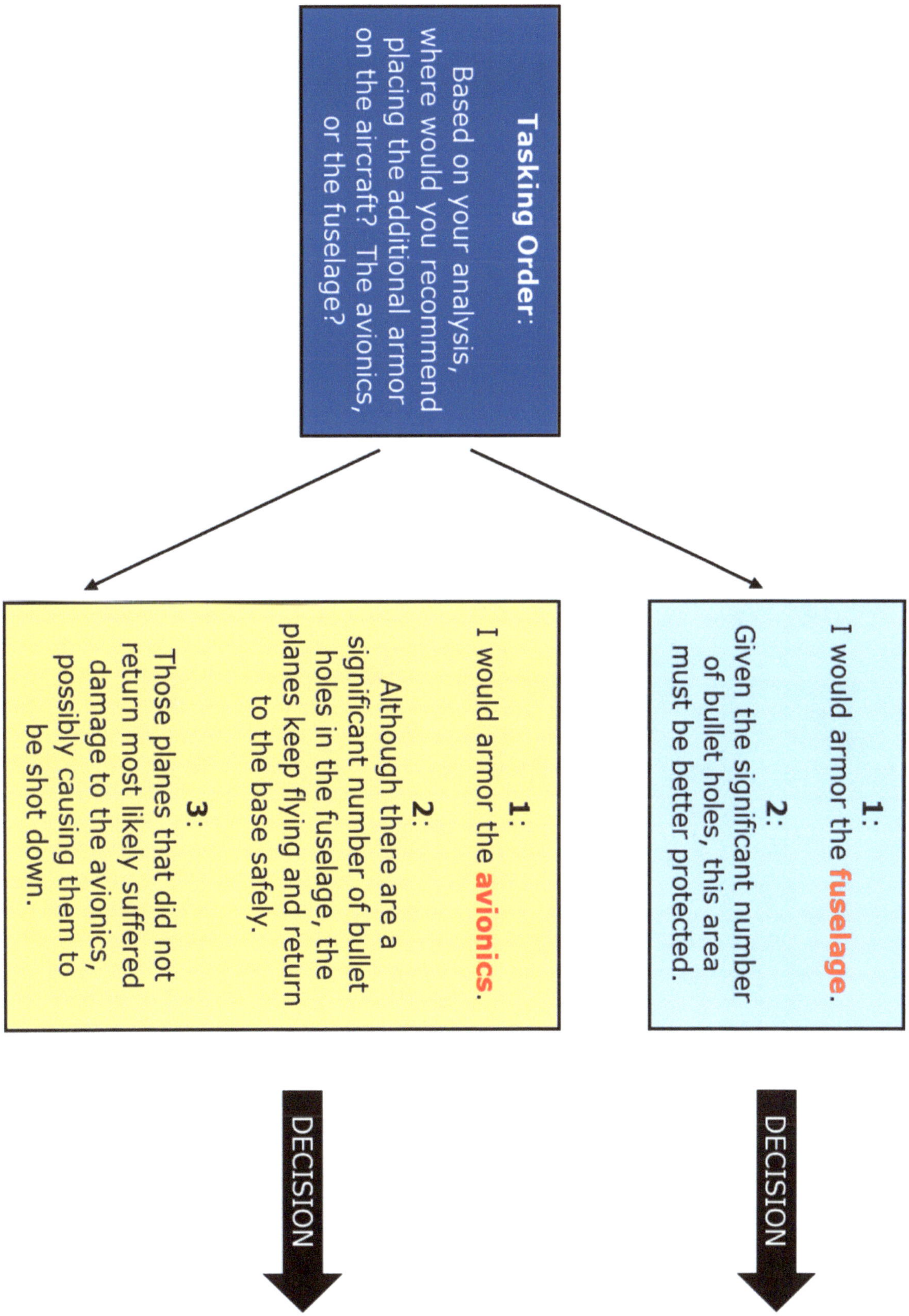

Image 5.2: Collection Requirements Management Process

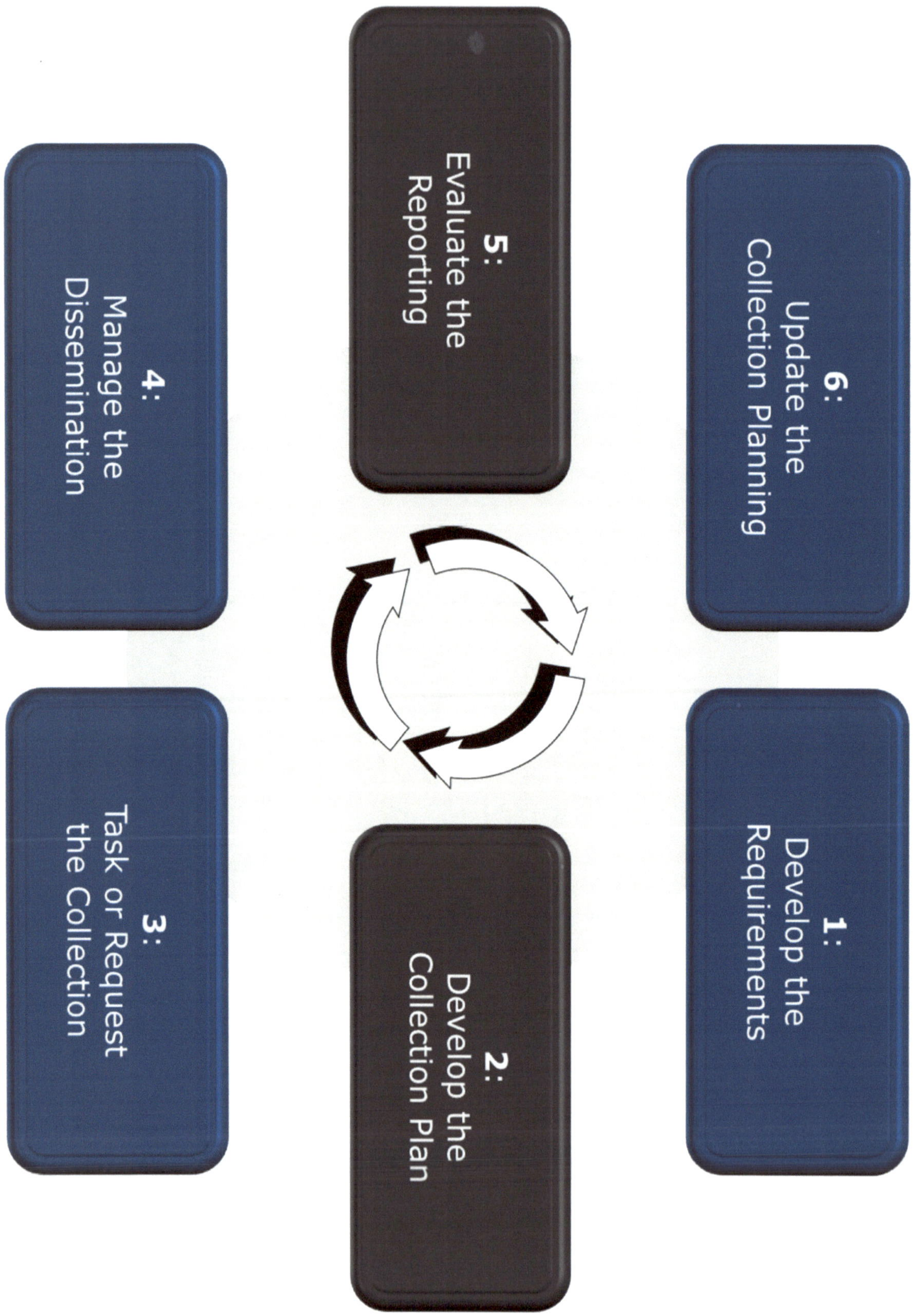

Image 6.1: Overview: ASCOPE / PMESII Matrix

ENVIRONMENTAL FACTORS / OPERATIONAL VALUES

	P Political	**M** Military	**E** Economic	**S** Social	**I** Information	**I** Infrastructure
A Areas	Areas Political	Areas Military	Areas Economic	Areas Social	Areas Information	Areas Infrastructure
S Structures	Structures Political	Structures Military	Structures Economic	Structures Social	Structures Information	Structures Infrastructure
C Capabilities	Capabilities Political	Capabilities Military	Capabilities Economic	Capabilities Social	Capabilities Information	Capabilities Infrastructure
O Operations	Operations Political	Operations Military	Operations Economic	Operations Social	Operations Information	Operations Infrastructure
P People	People Political	People Military	People Economic	People Social	People Information	People Infrastructure
E Events	Events Political	Events Military	Events Economic	Events Social	Events Information	Events Infrastructure

Image 6.2: Example: ASCOPE / PMESII Matrix for APT33

ENVIRONMENTAL FACTORS

OPERATIONAL VALUES

	P Political	M Military	E Economic	S Social	I Information	I Infrastructure
A Areas	Regional Boundaries	Offensive, Defensive, and Current Operations		Dark Web, BBS, and Chat	Social Media and the Dark Web	Internet Service Providers
S Structures			Crypto Currency Exchanges			
C Capabilities	Local and National Political Influence	Strengths, Weaknesses, Opportunities, and Threats	Access to Funding Sources	Direct and Local Global (WEB)	Education, Training, and Technology	Ability to Maintain and Service Technology
O Organizations	Criminal, Legal, Political	Governmental and Nonstate	State Sponsored and Criminal	The Black Hat Consortium	TV Media and News Groups	Local Tech Companies
P People	Government and Provincial Leaders	Local, Proxy, National, and International				Programmers, Developers, Hardware, and HUMINT
E Events	Rallies, Protests, and Meetings	Exploit Events and Operations			Propaganda Recruiting Fundraising Campaigns	

Image 6.3: Example Issue Decomposition: Cyber Threat

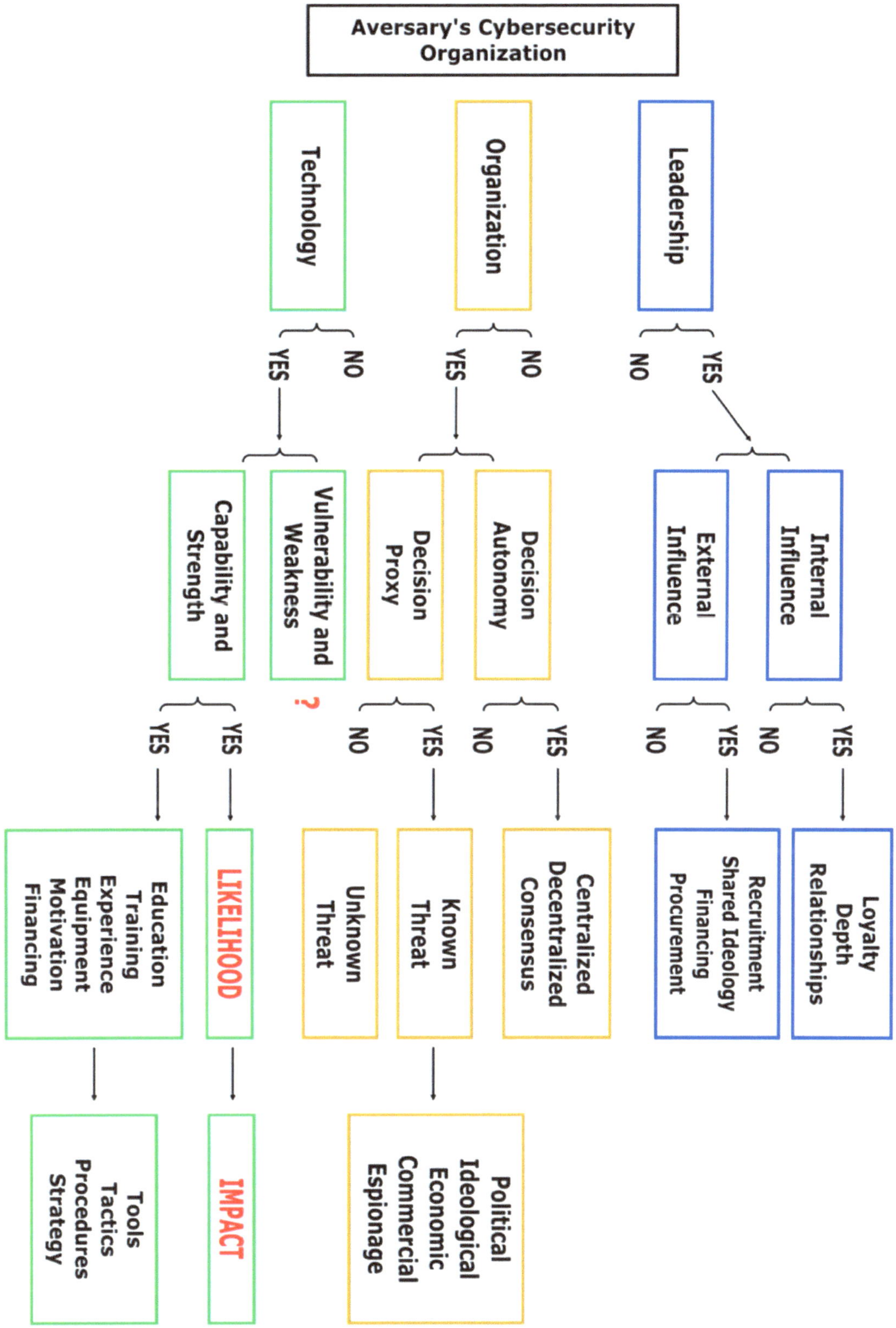

Image 7.1: The Model Hierarchy

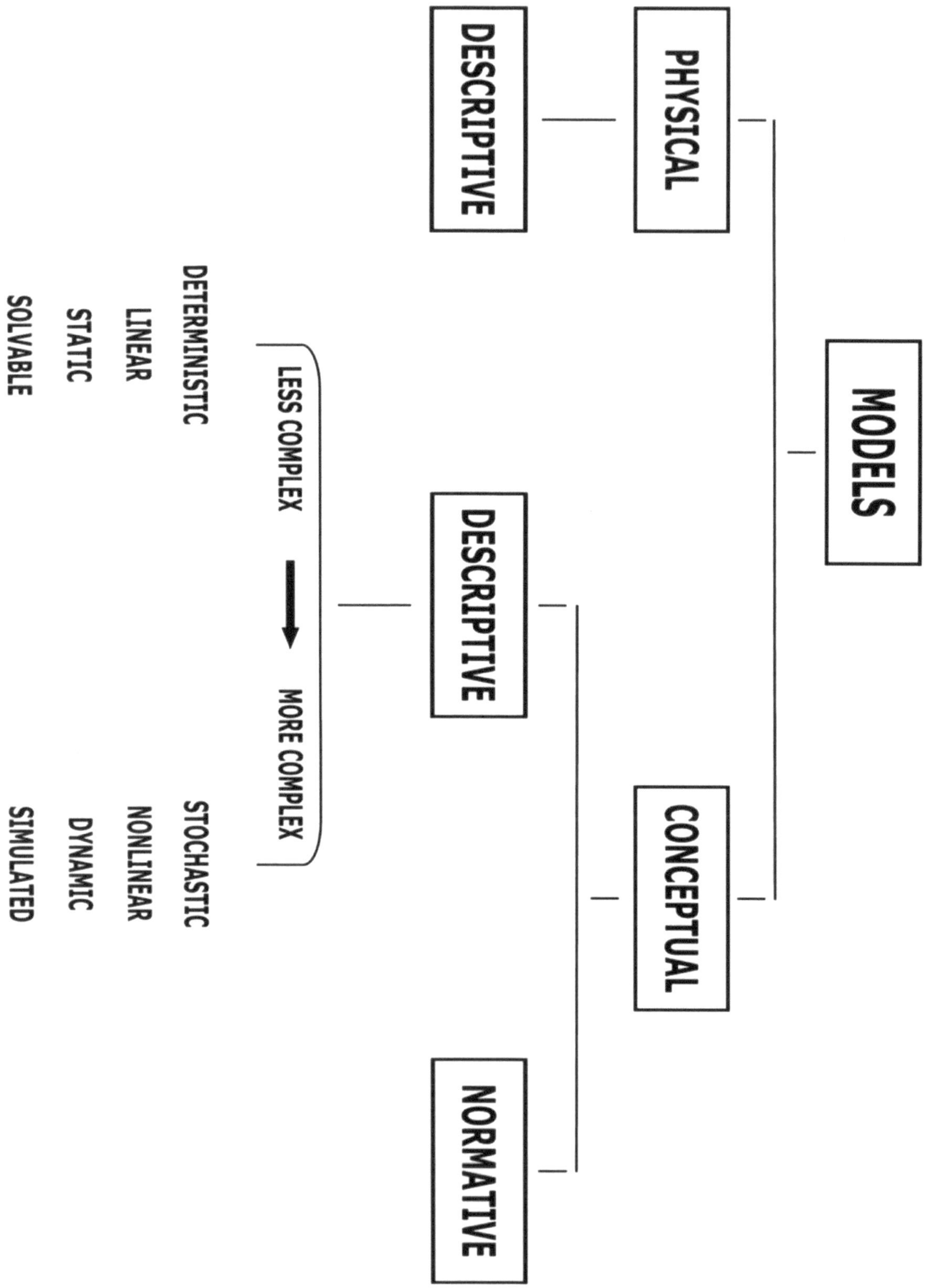

Image 7.2: Example: APT Comparative Threat Matrix

Threat Matrix: APT Comparative Analysis

Risk Values 1-20	**Known TTPs** Documented	**APT 28** Fancy Bear	**APT 32** OceanLotus	**APT 33** Elfin Team	**APT 41** Wicked Panda
1-20	Targets Petrochemical and Energy Industries	0	0	20	1
1-20	Targets Regions of Organizational Operation	0	0	20	1
1-20	Capable of Executing Zero-Day Exploits	2	4	5	6
1-20	Uses Boot or Auto-Start Execution	15	15	10	10
1-20	Targets Non-Standard Ports and/or Protocols	15	15	10	1
1-20	Uses Encryption for C2 and Data Exfiltration	15	15	15	10
1-20	Targets Windows Servers and File Sharing Systems	15	15	15	10
1-20	Uses Spear Phishing to Compromise Accounts	20	20	20	7
Total Value 1-160		**82**	**84**	**115**	**46**

Small Number Documented	Moderate Number Documented	Significant Number Documented
	Risk Values	
1-4 / **Low Probability**	5-10 / **Moderate Probability**	11-20 / **High Probability**

Image 7.3: Network Target Framework: Top Level

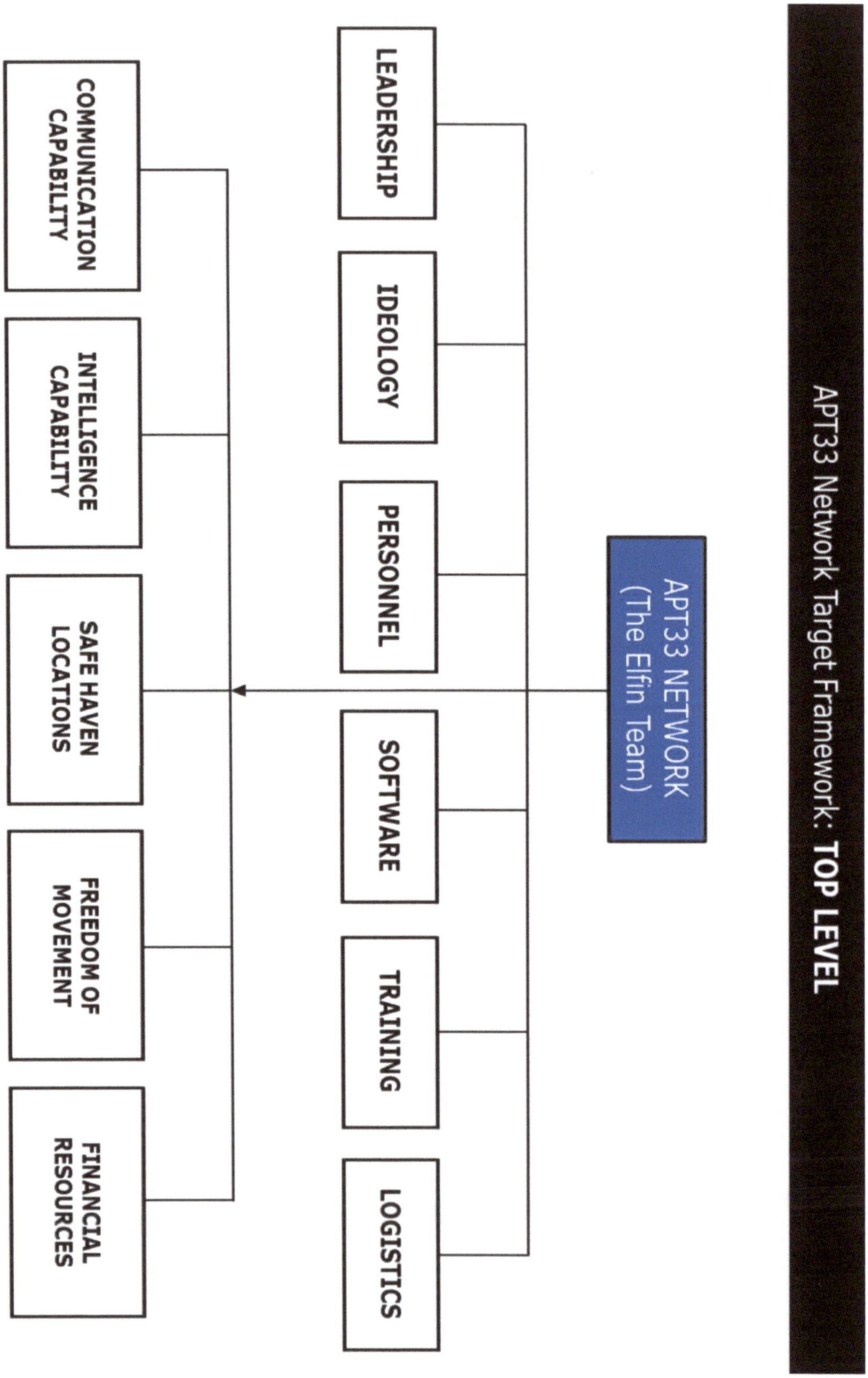

Image 7.4: Network Target Framework: Deconstruction 01

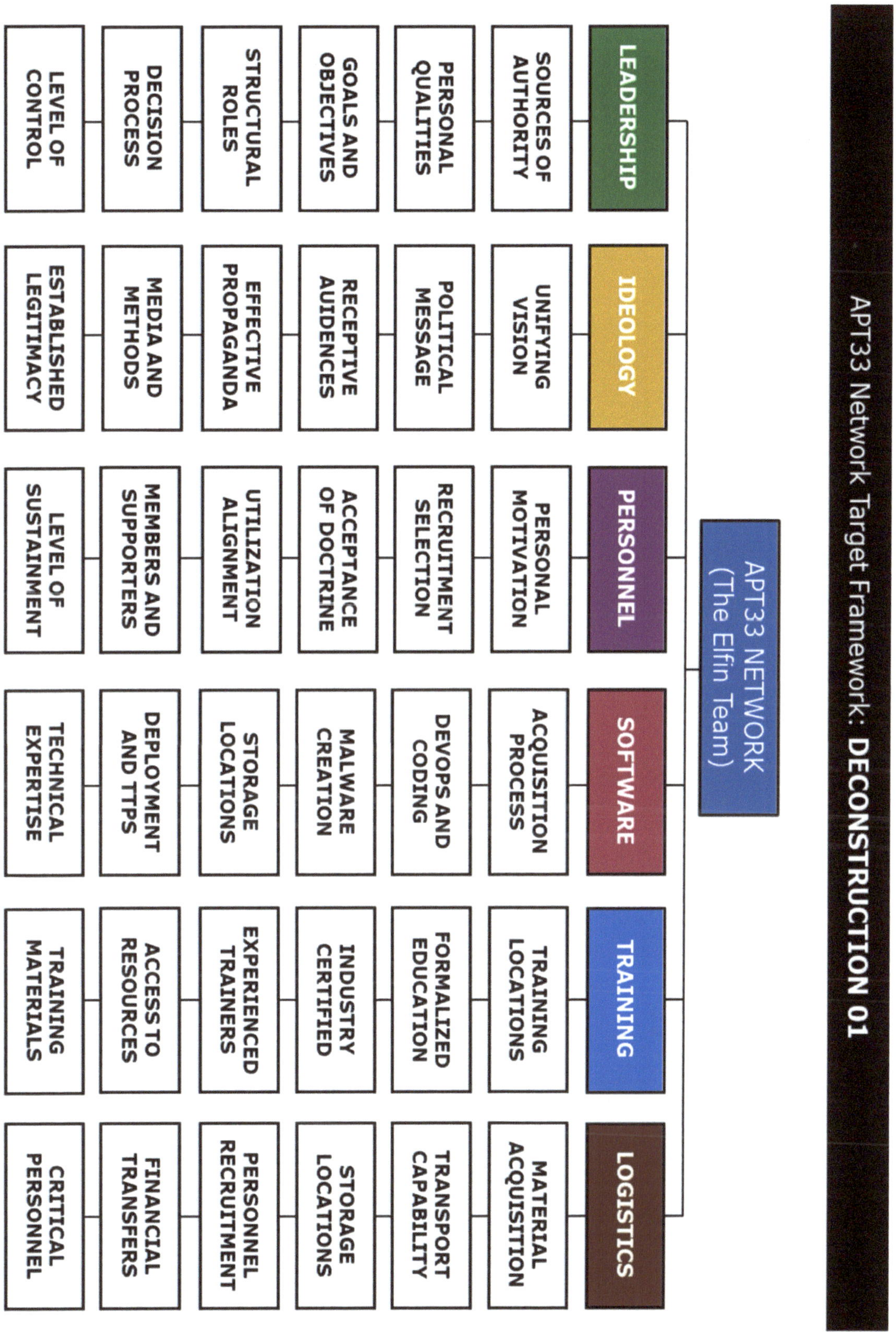

Image 7.5: Network Target Framework: Deconstruction 02

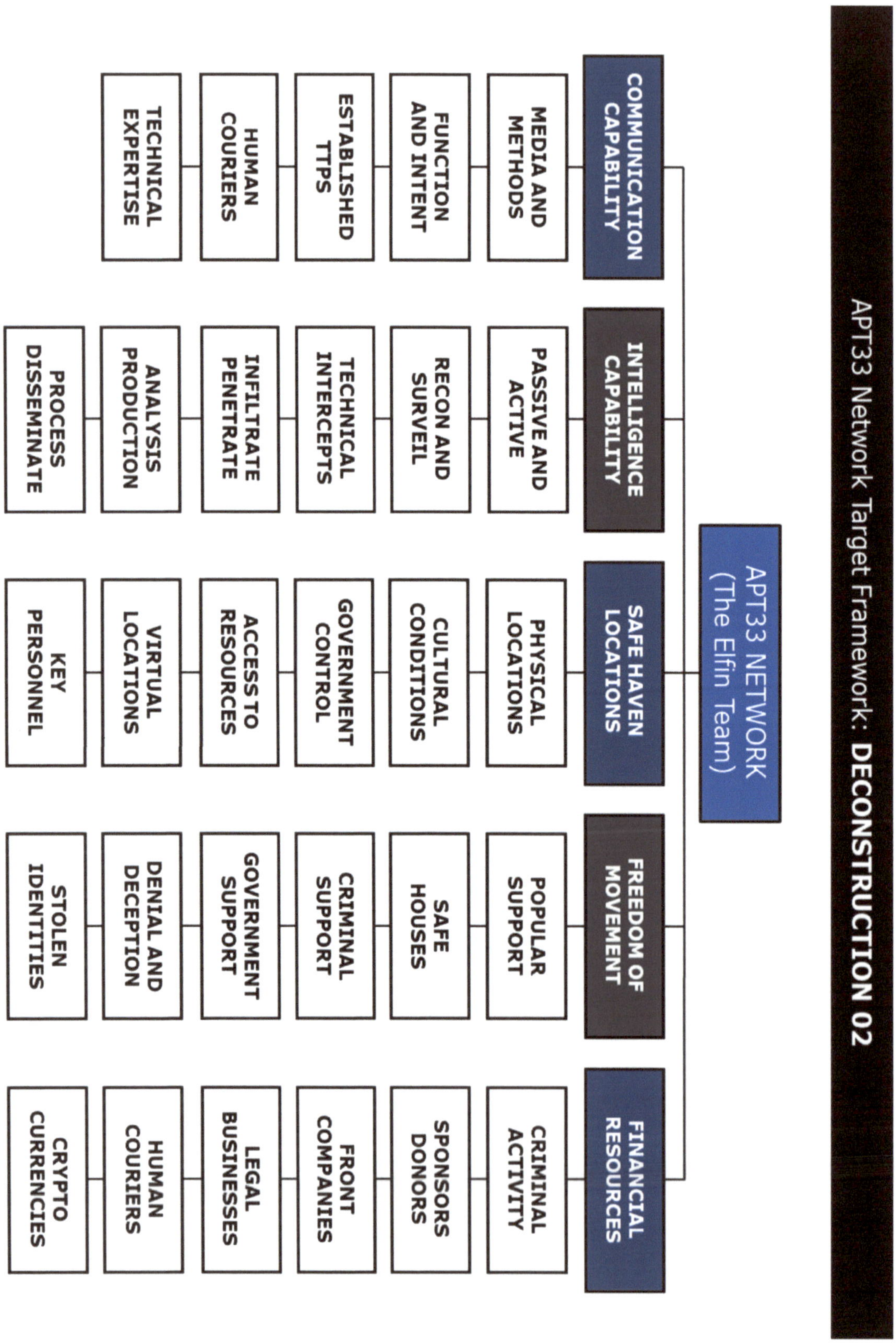

Image 9.1: The Estimative Methodology

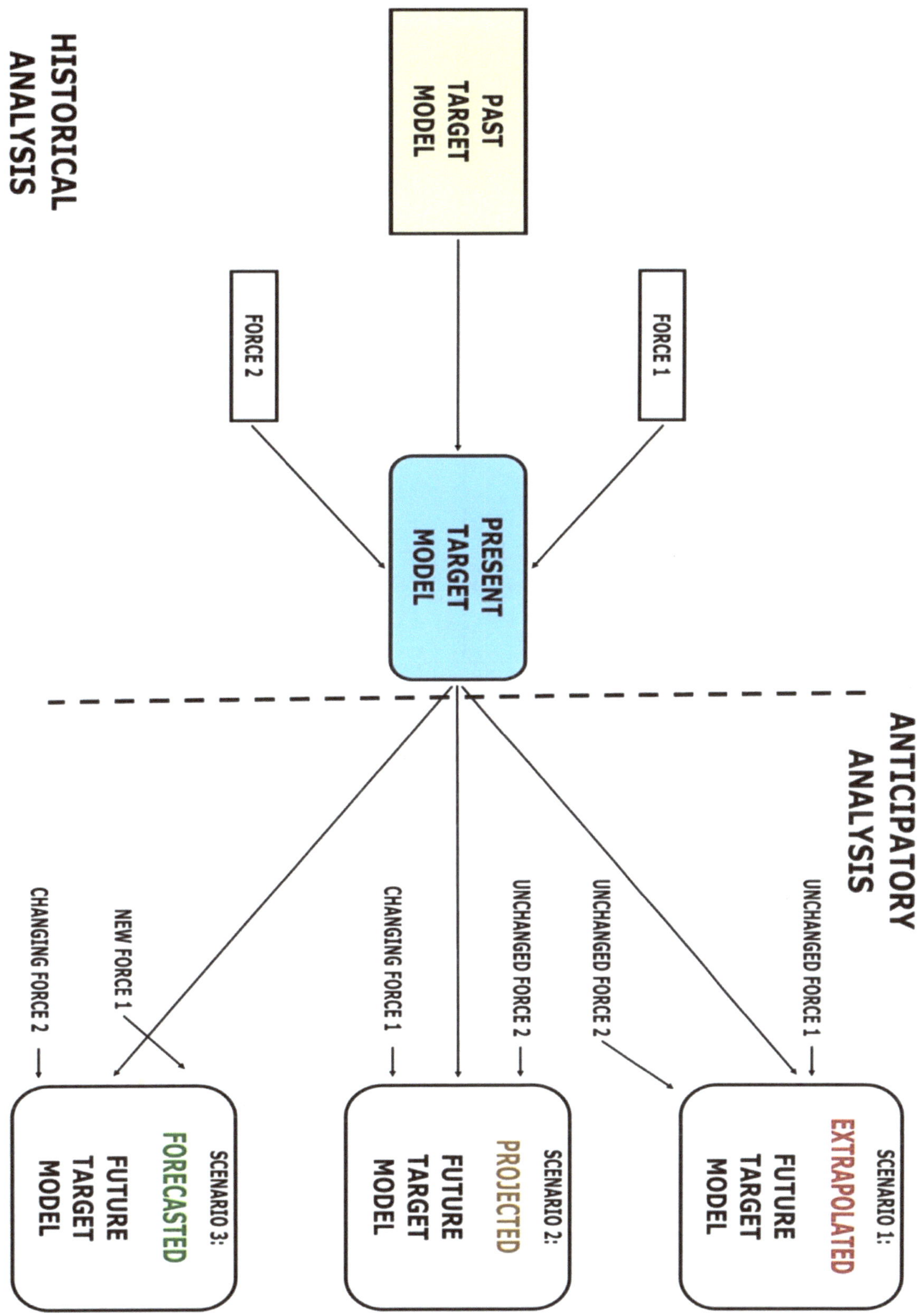

Image 9.2: Example: Influence Tree Diagram

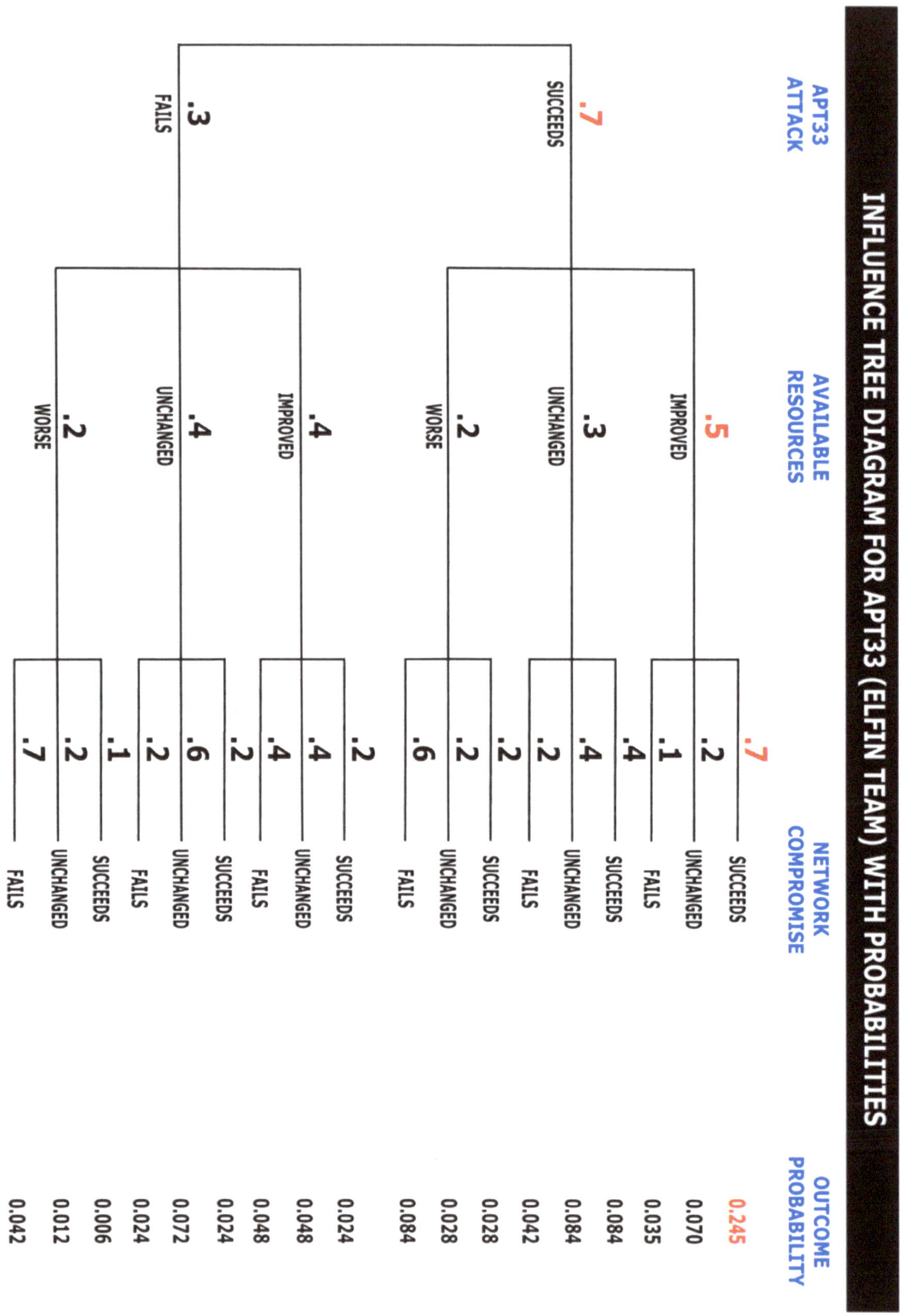

Image 9.3: Example: Influence Net Diagram

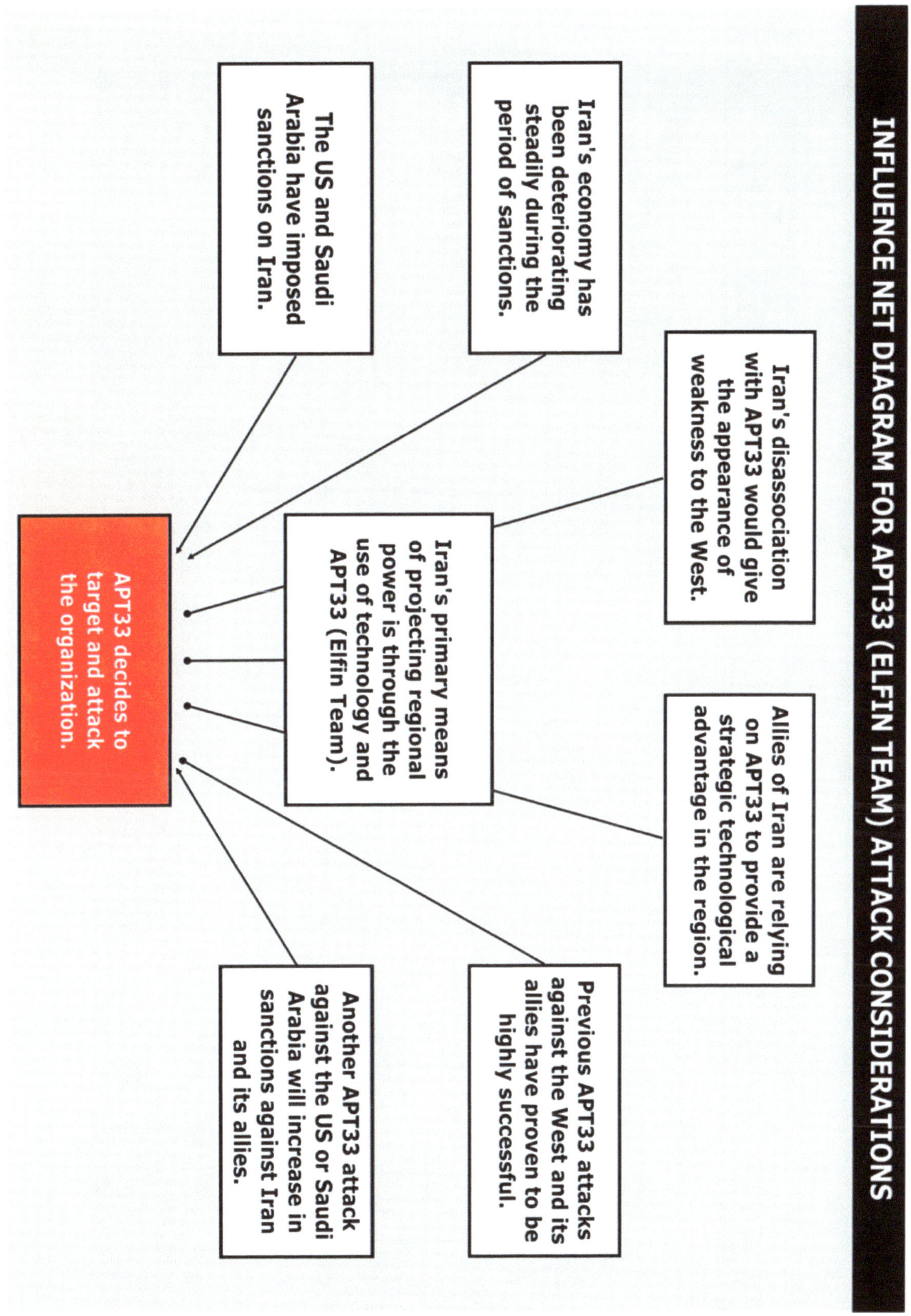

Image 10.1: Matrices: Example

ENVIRONMENTAL FACTORS

OPERATIONAL VALUES

	P Political	M Military	E Economic	S Social	I Information	I Infrastructure
A Areas	Areas Political	Areas Military	Areas Economic	Areas Social	Areas Information	Areas Infrastructure
S Structures	Structures Political	Structures Military	Structures Economic	Structures Social	Structures Information	Structures Infrastructure
C Capabilities	Capabilities Political	Capabilities Military	Capabilities Economic	Capabilities Social	Capabilities Information	Capabilities Infrastructure
O Operations	Operations Political	Operations Military	Operations Economic	Operations Social	Operations Information	Operations Infrastructure
P People	People Political	People Military	People Economic	People Social	People Information	People Infrastructure
E Events	Events Political	Events Military	Events Economic	Events Social	Events Information	Events Infrastructure

Image 10.2: Matrices: Example

Options	Objectives	Benefit	Risk	Implications	Indication
Organization's Public-Facing Website	Remove the ability to interact with the general public	Show of power and an embarrassment to the company	Mission failure, public embarrassment, and risk of sanctions	APT33 gains stature, Iran gains influence, new global sanctions	Ping scans, ping sweeps, and signs of potential DDoS
Organization's ICS / SCADA Control System	Destroy the ability to conduct critical operations	Show of power, media attention, and aid recruitment	Adequate training and maintaining security during the attack	APT33 gains stature, new sanctions, energy shortage	Increased phishing, actions consistent with ATT&CK Matrix
Organization's IAM / Password Database	Compromise the integrity of the credentialing system	Long-term access to sensitive data and accounts	Adequate training and maintaining security during the attack	APT33 gains stature, disrupt security, impact operations	Signs of network enumeration and increased phishing
Organization's Payroll / Finance Database	Disrupt the ability to conduct vital financial transactions	Show of power, disrupt operations and employee morale	Adequate training and maintaining security during the attack	APT33 gains stature, disrupt operations, loss of employees	Signs of network enumeration and spear phishing
Organization's Human Resource Database	Compromise and/or destroy employee personnel data	Acquire data on key employees for targeting efforts	Adequate training and maintaining security during the attack	APT33 gains stature, disrupt security, impact operations	Signs of network enumeration and spear phishing
Organization's Client Services Database	Compromise and/or destroy client business data	Show of power, disrupt operations and client confidence	Adequate training, security during attack and client response	APT33 gains stature, new sanctions, local economy impacted	Signs of network enumeration and attacks on clients
Organization's User Endpoint Devices	Establish a foothold for a large-scale network compromise	Acquire data about users and daily operations	Mission failure, public embarrassment, and impact recruitment	APT33 gains stature, disrupt security, impact operations	Increased phishing, event-based alerting, and help desk calls

Image 10.3: Example Event Mapping: APT33 Attack

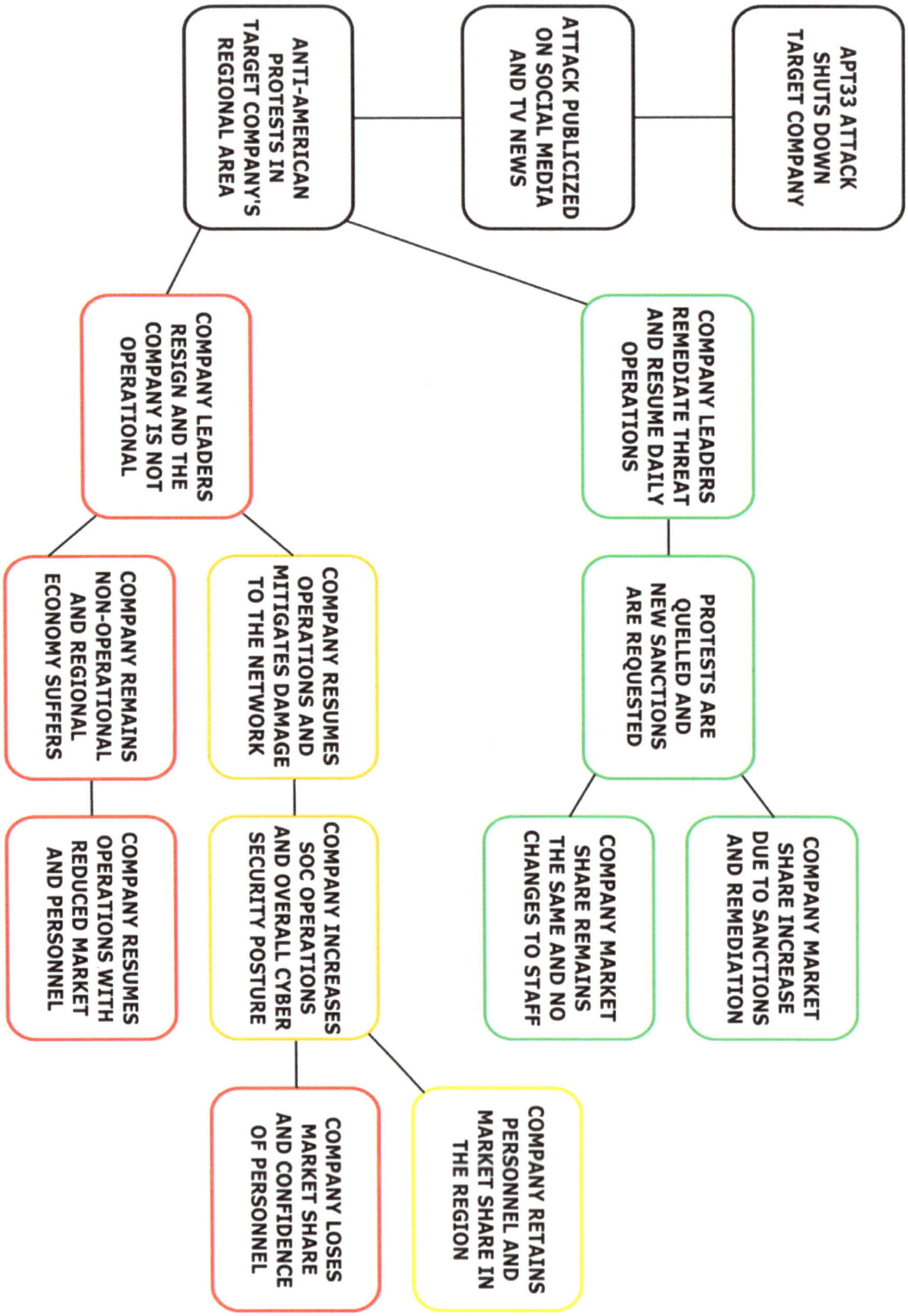

Image 10.4: Example Event Tree: APT33 Attack

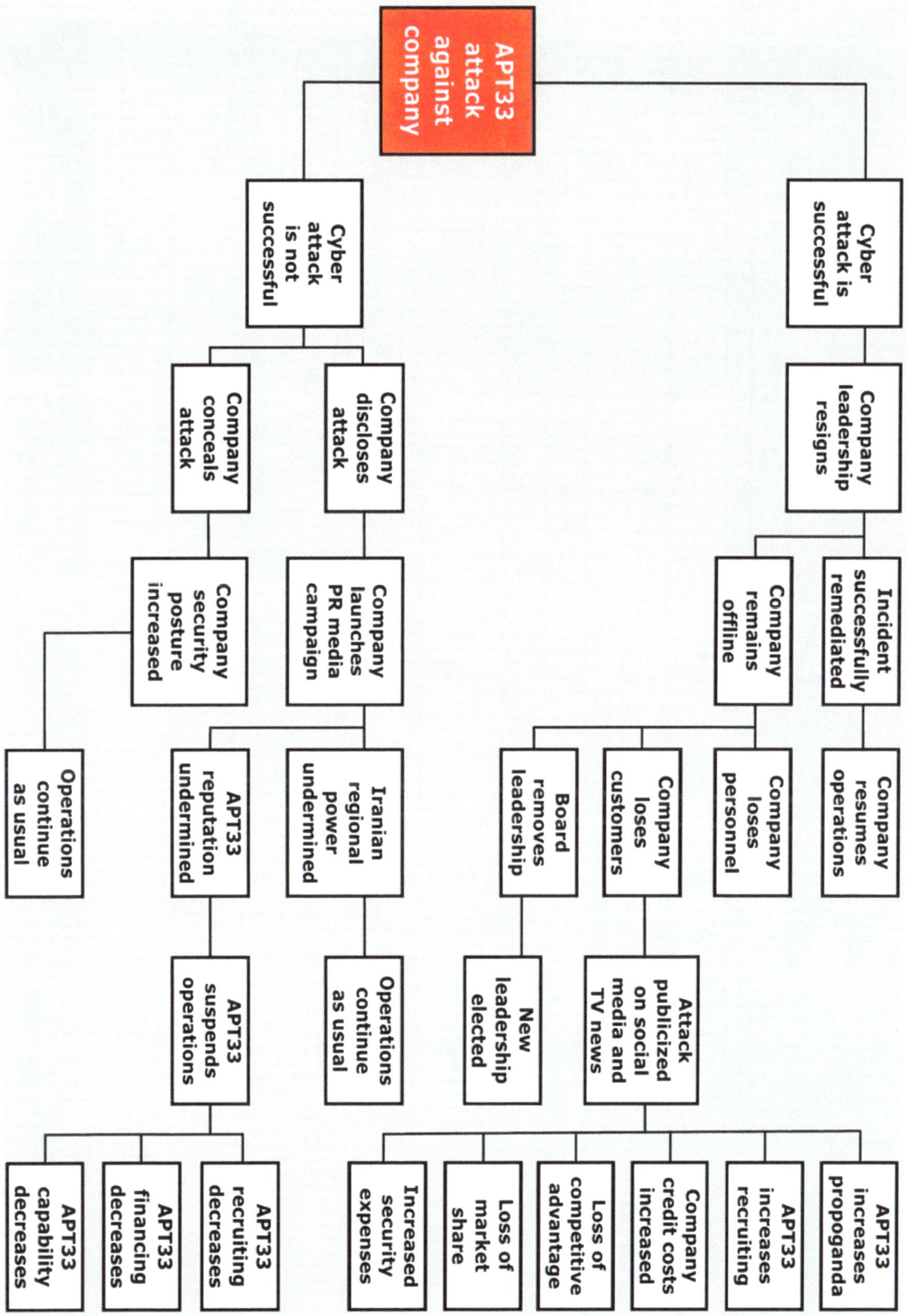

Image 10.5: Subjective Probability: Sample Table

Subjective Probability Table		
Term	Score	Range (Percentage)
Highly Probable	18-20	91 to 100
Probable	16-17	81 to 90
Highly Likely	11-15	71 to 80
Likely	8-10	61 to 70
Possible	5-7	41 to 60
Unlikely	4	31 to 40
Highly Unlikely	3	21 to 30
Improbable	2	11 to 20
Highly Improbable	1	1 to 10

Image 10.6: Subjective Probability: Methodology (I)

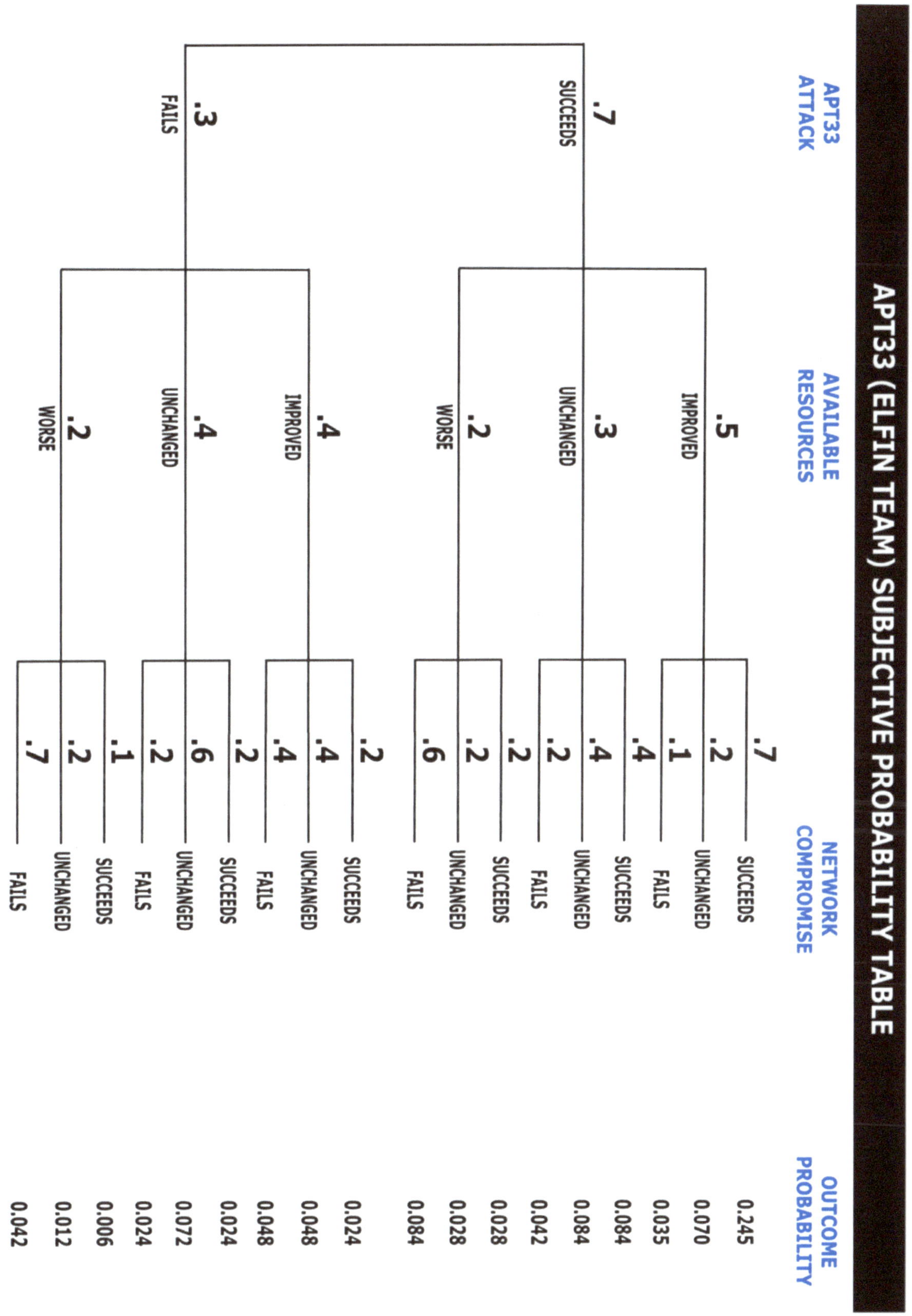

Image 10.7: Subjective Probability: Methodology (II)

Threat Matrix: APT Comparative Analysis

Risk Values 1-20	**Known TTPs** Documented	**APT 28** Fancy Bear	**APT 32** OceanLotus	**APT 33** Elfin Team	**APT 41** Wicked Panda
1-20	Targets Petrochemical and Energy Industries	0	0	20	1
1-20	Targets Regions of Organizational Operation	0	0	20	1
1-20	Capable of Executing Zero-Day Exploits	2	4	5	6
1-20	Uses Boot or Auto-Start Execution	15	15	10	10
1-20	Targets Non-Standard Ports and/or Protocols	15	15	10	1
1-20	Uses Encryption for C2 and Data Exfiltration	15	15	15	10
1-20	Targets Windows Servers and File Sharing Systems	15	15	15	10
1-20	Uses Spear Phishing to Compromise Accounts	20	20	20	7
Total Value 1-160		**82**	**84**	**115**	**46**

Small Number Documented	Moderate Number Documented	Significant Number Documented
Risk Values		
1-4 / **Low Probability**	5-10 / **Moderate Probability**	11-20 / **High Probability**

www.ingramcontent.com/pod-product-compliance
Lightning Source LLC
LaVergne TN
LVHW070946210926
848482LV00002B/106